Naturalizing Heidegger

SUNY series in Environmental Philosophy and Ethics
―――――――
J. Baird Callicott and John van Buren, editors

Naturalizing Heidegger

*His Confrontation with Nietzsche,
His Contributions to Environmental Philosophy*

David E. Storey

Published by State University of New York Press, Albany

© 2015 State University of New York

All rights reserved

Printed in the United States of America

No part of this book may be used or reproduced in any manner whatsoever without written permission. No part of this book may be stored in a retrieval system or transmitted in any form or by any means including electronic, electrostatic, magnetic tape, mechanical, photocopying, recording, or otherwise without the prior permission in writing of the publisher.

For information, contact State University of New York Press, Albany, NY
www.sunypress.edu

Production, Eileen Nizer
Marketing, Anne M. Valentine

Library of Congress Cataloging-in-Publication Data

Storey, David E., 1982–
  Naturalizing Heidegger : his confrontation with Nietzsche, his contributions to environmental philosophy / David E. Storey.
       pages cm. — (SUNY series in environmental philosophy and ethics)
  Includes bibliographical references and index.
   ISBN 978-1-4384-5483-2 (hc : alk. paper)—978-1-4384-5482-5 (pb : alk. paper)
   ISBN 978-1-4384-5484-9 (ebook)
    1. Heidegger, Martin, 1889–1976. 2. Environmental ethics. 3. Philosophy of nature. 4. Environmental sciences—Philosophy. 5. Nietzsche, Friedrich Wilhelm, 1844-1900.   I. Title.

B3279.H49S755 2015
193—dc23   2014008387

10 9 8 7 6 5 4 3 2 1

Contents

Acknowledgments	vii
Introduction	1
Chapter 1 The Traditional Reading of Heidegger's Relevance for Environmental Philosophy and Ethics	11
Chapter 2 The Question concerning Biology: Life, Soul, and Nature in Heidegger's Early Aristotle Lecture Courses	33
Chapter 3 Life and Nature in *Being and Time*	47
Chapter 4 Back to Life: Organism, Animal, and *Umwelt* in *Fundamental Concepts of Metaphysics*	81
Chapter 5 Nature in the Later Heidegger: Earth, *Physis*, Technology, Machination, and Poetic Dwelling	107
Chapter 6 Nature and Nihilism: Heidegger's Confrontation with Nietzsche	141
Chapter 7 Naturalizing Nietzsche: Life, Evolution, and Value	173

Chapter 8
Engaging Environmental Ethics   217

Notes   235

Bibliography   261

Index   271

Acknowledgments

I want to thank the editors and reviewers at SUNY Press for publishing my work and for their feedback and support in bringing this project to completion. I am grateful to the following publishers for granting permission for material to be republished here. A version of chapter 2 appeared as "Heidegger and the Question Concerning Biology: Life, Soul, and Nature in the Early Aristotle Lecture Courses," in *Epoché: a Journal for the History of Philosophy* 18, no. 1 (Fall 2013). A version of chapter 7 appeared as "Nietzsche's Non-Reductive Naturalism: Evolution, Teleology, and Value," in *Pli: The Warwick Journal of Philosophy* 23 (2012): 128S–52. A small portion of chapter 7 will appear in the journal *Philosophy in Review* in a book review of *Nietzsche, Naturalism, and Normativity*, edited by Christopher Janaway and Simon Robertson.

Heartfelt thanks go to Professor John Van Buren for his guidance and mentorship over the years, for helping me become a scholar, and for going above and beyond the call of duty to help me bring this project to completion. I am deeply indebted to Professor Michael E. Zimmerman, whose work shaped my own and who supported me in this project before we had even met. I also want to thank Paul McNellis, S.J., for first introducing me to philosophy; the philosophy departments at Fordham University and Boston College; my colleagues and friends Eleanor Helms, David Zoller, Adam Konopka, and Shane Wilkins, for being such excellent and supportive conversation partners throughout the years; my old friends Brian Johnson, Patrick Lazarus, Chris King, John Haley, Sean Knight, John Andiola, Jonathan Steets, Kyle Corigliano, Eliza Bent, Tania Rudnitski, and Chris Sterback for keeping me grounded and sane; Sal Giambanco, for unexpected inspiration; and Ronald Tacelli, S.J., for believing in me when it mattered most. Finally, I want to thank my parents, Francis and Susan, for their boundless love and support.

Introduction

Naturalizing *Heidegger*? Given that Heidegger is a phenomenologist, and that phenomenology is traditionally opposed to naturalism, any attempt to naturalize Heidegger's thought might at first seem, to use his own phrase, "a round square and a misunderstanding." But it all depends on what we mean by naturalism.

My argument in this book is that resources from the work of Martin Heidegger and Friedrich Nietzsche can help us build a nonreductive naturalism that can support an environmental ethic. But what is a "nonreductive" naturalism? Why should it support an environmental ethic? Why Heidegger? Why Nietzsche? Why both of them together?

To answer these questions, I want to start with a more basic one: *Is value natural?* We could phrase the question in other and more precise ways: Are the values we ascribe to nature created or discovered? Do living things value or have value? Do animals? Does nature itself? Though it may be a crude form of the question, it is arguably the basic question of environmental philosophy, and it raises fundamental axiological and metaphysical questions that reach deep into the heart and history of Western thought.

While the question is old, it gains new significance in the modern world and new urgency today. One of the mainsprings of modernity is the emergence of a new conception of nature in the seventeenth century. In the most general sense, this shift from the medieval to the modern is the move from a teleological to a mechanistic conception of nature. In his magisterial study, *The Metaphysical Foundations of Modern Science*, E. A. Burtt conveys just how seismic this shift truly was:

> For the dominant trend in medieval thought, man occupied a more significant and determinative place in the universe than the realm of physical nature, while for the main current of modern thought, nature holds a more independent, more determinative,

> and more permanent place than man. . . . For the Middle Ages, man was in every sense the center of the universe. . . . The entire world of nature was held not only to exist for man's sake, but to be likewise immediately present and fully intelligible to his mind. Hence the categories in terms of which it was interpreted were not those of time, space, mass, energy, and the like; but substance, essence, matter, form, quality, quantity—categories developed in the attempt to throw into scientific form the facts and relations observed in man's unaided sense experience of the world and the main uses which he made it to serve.[1]

So long as the teleological conception held sway, be it the Greeks' eternal cosmos or the Christians' divine creation, the notion that there was a necessary connection between the normative and the natural was generally taken for granted. But with the advent of the mechanistic view, the strength and status of that connection was thrown into serious doubt and has been ever since.

One would have thought that the modern decentering of humanity would be a victory for nature. In a dance, when one partner becomes less self-centered and more aware of and attuned to the other's movements, the result tends to be better for both. However, it appears that the opposite occurred: the mechanistic worldview enabled humans to exert unprecedented control over nature. Why? Part of the answer may be that in the traditional view, nature was seen as subordinate to humans, but it was still taken to be good. In the modern view, nature is no longer subordinate, but it is no longer seen as good. Morally, the mechanistic universe is neutral and acentric, and so, given our tendencies to preserve and care for ourselves, our kin, and our kind, humans again became the center; the difference was that now, we were armed with new and more powerful tools to manipulate nature to conform to our own ends, whatever they happened to be. With the material constraints of technological immaturity, the moral constraints of something like natural law, and the mental constraints of a teleological vision of nature overcome, we were free to remake nature as we saw fit; and this appears to be more or less what we have done over the last three hundred years.

The basic question takes on new urgency today because we are surrounded by the results of organizing human affairs according to this new conception of nature: the myriad messy health, energy, economic, ecological, atmospheric, moral, and perhaps even spiritual problems we call the environmental crisis. The increasingly conventional wisdom is that to address the environmental crisis we need major changes in personal lifestyles, professional practices, and public policies, and it is often said that these changes demand a change in values. But the challenge to rethink our values entails rethinking the worldview from which they flow. It entails, in other words, rethinking our view of nature and our place within it, and this means reckoning with metaphysics. And it is safe to say that the regnant metaphysical framework of our age is something called "naturalism."

Naturalism is said in many ways. It could mean "methodological" naturalism, which eschews metaphysical claims and merely purports to study "empirical" phenomena, things that can be observed with the senses or their instrumental extension. It could mean "metaphysical" naturalism, which makes the stronger claim that only such phenomena are real. According to Keith Campbell, "metaphysical naturalism affirms that the natural world is the only real one, and that the human race is not separate from it, but belongs to it as a part. . . . The natural world is the world of space, time, matter, energy, and causality."[2] Though naturalism, in this broad sense of materialism, is nothing new—Democritus and Lucretius were, after all, naturalists of a kind—naturalism has become our default setting mainly because of its association with the conceptual and empirical foundations of science. In a recent anthology devoted to the theme of naturalism, Mario De Caro and David Macarthur label this view "scientific naturalism": "[S]cientific naturalists typically conceive nature as a causally closed spatio-temporal structure governed by efficient causal laws—where causes are thought of, paradigmatically, as mind-independent bringers-about of change or difference."[3] As they point out, scientific naturalists not only "claim that the conception of nature of the natural sciences is very likely to be true," but they go further by insisting that "this is our only bona fide or unproblematic conception of nature."[4] In this way, scientific naturalism is one of the most entrenched default settings in contemporary philosophical discourse.

However, there is widespread consensus, among both continental and analytic philosophers, that scientific naturalism is very problematic indeed. Beginning with Husserl's critique of psychologism, phenomenology has largely defined itself in opposition to naturalism in maintaining that it cannot account for meaning and intentionality and is essentially an incoherent epistemological and metaphysical position. Likewise, philosophers of mind, biology, science, epistemology, metaphysics, and ethics associated with the analytic tradition have begun to challenge the basic premises of naturalism.[5] One of the chief objections is that naturalism tends to be *reductive*: it purports to explain the higher in terms of the lower or one kind of beings in terms of a different kind of beings, but by doing so, it distorts the higher or erases the qualitative differences between beings. Nevertheless, it is an enduring temptation, partly because scientific thinking has become such an integral part of our intellectual culture, and partly because its chief traditional rivals, dualism and hylomorphism, are difficult to defend; the former because of problems of the existence of immaterial substances and their interactions with the material world, and the latter because it seems wedded to the teleological conception of nature science is widely seen as having displaced. In this sense, we might apply Churchill's quip about democracy to naturalism: it is the worst metaphysical view we have, except for all the others. But the starting point for challenging it is the recognition that it is simply one metaphysical position among others.

Moreover, aside from theoretical problems facing naturalism, there are its practical results. Indeed, one of the chief concerns of many environmental ethicists is to recapture a vision of the natural world infused with value; but this concern can only arise in a context in which nature is not afforded any kind of moral status. Put another way, the modern conception of nature is nihilistic. Nihilism not only drains the world of purpose for humans. It also entails that animals, living beings, and nature itself have no value worthy of respect and moral consideration. And it issues from the modern conception of nature. Alfred North Whitehead's formulation captures the connection: "Nature is a dull affair, soundless, scentless, colourless; merely the hurrying of material, endlessly, *meaninglessly*."[6] Bertrand Russell was more dramatic: "Brief and powerless is man's life. On him and all his race the slow, sure doom falls pitiless and dark. Blind to good and evil, reckless of destruction, omnipotent matter rolls on its relentless way."[7] But in response to this ultimate impotence, human beings have come to exercise greater power over nature than ever before, trying to attain complete security in an entropic universe. There is thus a common link between existentialism and environmentalism: if God does not exist, so the saying goes, then everything is permissible; if values are purely anthropogenic and lack cosmic support, everything is permissible with respect to the environment.[8]

There is, then, considered opposition to scientific naturalism fueled by the recognition not only that its practical, civilizational consequences are bad, but that its theoretical basis is unsound, or at least eminently questionable. Yet if dualism is indeed dead, if we cannot appeal to the supernatural, then we must alter what we mean by the natural. This is why I think a nonreductive naturalism and an environmental ethic go hand in hand: the destruction of the natural environment in large part depends on reducing nature to the sum of material objects capable of mathematical description and manipulation by human artifice, and a solid case for curtailing such destruction can be built on the basis of a nonreductive conception of nature. But what kind of naturalism would qualify as nonreductive? Ted Benton sketches the guiding idea:

> A non-reductionist naturalism, making use of the ideas of a hierarchy of more or less autonomous levels of organization of matter, each with its own, qualitatively new, 'emergent' powers or properties has been one fruitful way of maintaining the insights of a naturalistic approach, without falling foul of what is valid in the anti-naturalistic critique. Such hierarchical, 'emergent powers' ontologies enable their advocates to recognize in the various subject matters of the different natural and social sciences more or less discrete and autonomous object-domains, while at the same time making no concessions to spiritualistic, vitalist, or supernatural beliefs.[9]

To that end, in this book I approach the question of natural value by laying the groundwork for a nonreductive naturalism, and do so by drawing on the work of Heidegger and Nietzsche. Why these two thinkers? For one, they both acutely perceived that the modern conception of nature led to nihilism. Heidegger's use of phenomenology to return to the question of being led him to resist scientific naturalism and explore alternative conceptions of nature throughout his work. Nature is no peripheral concern in his thought; indeed, in his later work, being—his chief quarry—becomes all but identical with *physis*, the Greek term for nature. Nietzsche, writing shortly after Darwin's discovery of evolution by natural selection, labored to work out a conception of nature that incorporated evolutionary theory but rejected the mechanistic metaphysics with which it was readily paired. Both thinkers, in different ways and to different degrees, embraced an alternative naturalism at some point in their work. And the naturalism to which their reflections point, I argue, provides us with a metaphysical foundation for an environmental ethic.

Over the last several decades, a number of scholars have attempted to extract an environmental ethic from Heidegger's thought, citing his critiques of science and technology (the early Michael E. Zimmerman), his philosophy of language (Charles Taylor), or his concept of poetic dwelling (Bruce V. Foltz). These attempts have mainly looked to Heidegger's later philosophy of nature as a resource for environmental ethics. I submit that this approach faces four problems: 1) Heidegger fails to develop a robust philosophical biology and deal with the "mind/life" problem; 2) he ignores evolutionary theory; 3) he rejects value theory; and 4) he regards nature as ultimately unintelligible.

At the same time, I contend that there are resources in his earlier work that point toward a different approach to environmental ethics: his early writings on Aristotle, his engagement with biologist Jakob von Uexküll, and especially his confrontation with Nietzsche. In Aristotle's *De Anima*, Heidegger finds a path toward establishing the intentionality of organisms and fitting humans along a common continuum with other living things. From Uexküll, he adopts a novel conception of the organism's relation to its environment that resists mechanism in biology, suggests a deep kinship between animals and humans, and prefigures his later conception of being-in-the-world. Through Nietzsche, he sees the attempt to overcome nihilism through a life-affirming vision of humanity's place in nature and a recognition of the organic basis of value. The unifying theme of these three threads is Heidegger's pioneering (though aborted) investigation into the being of life. In laying out Heidegger's nascent naturalism and ontology of life, I chart the evolution of his philosophy of nature from the early 1920s to his later work in the 1940s.

In chapter 1, I begin by surveying the numerous attempts to frame Heidegger as a protoecological thinker, which tend to cite themes in his later work as supportive of an environmental ethic. Next, I submit that these approaches, and Heidegger's later philosophy of nature, face the problems mentioned above. Third, I describe the project of naturalizing phenomenology and explain the difficulties of naturalizing Heidegger's thought. There are two phenomenological traditions—"transcendental" and "biological"—in tension in Heidegger's earlier work: the former is antinaturalist, while the latter points to a nonreductive naturalism. While he flirted with the second path, Heidegger ultimately chose the first.[10] I argue that there is an overlooked strain of naturalism in Heidegger's thought, running from his early work on Aristotle to his engagement with the biologist Jakob von Uexküll and, finally, to his confrontation with Nietzsche. These untapped resources suggest a *naturalized* Heidegger that provides a sounder basis for environmental ethics than the traditional approach to his thought.

In chapter 2, I trace Heidegger's investigations into life and animality in the lecture courses on Aristotle in the early–mid-1920s, focusing on 1) the influence of biologist Jakob von Uexküll on Heidegger's reading of Aristotle; 2) Heidegger's attraction to a "continuum" view of the human-animal and human-nature relationship; and 3) how the Aristotle lectures prefigure *Fundamental Concepts of Metaphysics*, his most sustained attempt at a philosophical biology. All told, Heidegger's early ontology of life paints a more continuous view of the human-nature relationship than his middle and later writings. Yet after the late 1920s, he never returns to this theme in great detail. His investigations into theoretical biology in the late 1920s could have led him to a nonreductive form of naturalism, rather than the poetic view of nature that dominates his later writings.

Chapter 3 examines the concept of nature in *Being and Time*. First, I analyze the text's aim, structure, and method. Second, I trace the emergence of the theme of nihilism in Heidegger's work in and before *Being and Time*. Third, I address his brief treatment of the concept of life in the text, noting how it departs from the views sketched in the previous chapter. Fourth, I discuss the three senses of nature addressed in the text, showing what misfires in his early approach to nature and how this prefigures his later, more anti-naturalist position. Heidegger's thinking about nature splits into two tracks after *Being and Time* that are never reconciled. One track leads to his 1929–30 lectures, *Fundamental Concepts of Metaphysics*, which addresses a host of issues pertinent to philosophical biology and ecology; this track leads in the direction of a naturalized phenomenology, a path Heidegger chose not to pursue. The second track leads to the anti-naturalist mainsprings of his later thought, such as the enigmatic notion of earth, his retrieval of *physis*, and his "history of being."

In chapter 4, I pursue the first track mentioned above, which culminates in *Fundamental Concepts of Metaphysics*. First, I return to the themes of nihilism (discussed in the previous chapter) and life (addressed in chapter 2). Heidegger's attempt to overcome nihilism after *Being and Time* is the context in which his investigations into life, biology, and animality take place: his strategy to combat nihilism is to recover a more adequate sense of nature. Second, I trace Heidegger's treatment of life, biology, and animality in *Fundamental Concepts of Metaphysics*. I submit that the moves toward a transcendental approach in *Being and Time* and toward poeticizing in his later thought constitute a wrong turn in his philosophy of nature. After 1930, Heidegger's thought takes a sharp antinaturalist turn.

Having detailed Heidegger's earlier approach to nature and life, I turn in chapter 5 to his later philosophy. I examine his post–mid-1930s approach to nature, detail how and why environmental thinkers have drawn on this aspect of his thought, and complete my argument that his philosophy of nature took a wrong turn after 1930. I focus on his unique notion of earth, his appropriation of *physis*, his notion of poetic dwelling, and his critiques of humanism, modern technology, and what he calls "machination." Many have seen Heidegger's turn as a turn towards nonanthropocentrism, or even biocentrism. I argue that this view is mistaken. Though several of

Heidegger's later concepts, such as *Gelassenheit*, are promising for environmental ethics, they are too vague and stray too far from the concrete realities of animal, biological, and natural phenomena. Despite the adoption of a poetic style, a "being-centric" standpoint, and the "holistic" tone in his later work, Heidegger's conviction about the ontological gulf between humans and nonhumans persists throughout the later work.

Picking up the trail of nihilism sketched in earlier chapters, in chapter 6 I examine Heidegger's view of nihilism as the logic of Western metaphysics, link it to his understanding of humanity's relation to nature, and compare it with Nietzsche's view of nihilism. At first, he sees Nietzsche's thought as the antidote to nihilism because of the latter's attacks on traditional metaphysics, idealism, and scientific naturalism, as well as Nietzsche's understanding of living being. Later on, however, Heidegger concludes that Nietzsche's thought—especially its call to naturalize and reanimalize human being—is the very essence of nihilism and must be overcome. I rebut Heidegger's interpretation of Nietzsche as an anthropocentric thinker and challenge his own view of nihilism.

In chapter 7, I show how ideas from Nietzsche's thought can supplement Heidegger's nonreductive naturalism. To be sure, there are problems with classifying Nietzsche as a nonreductive naturalist. Throughout his development, Nietzsche appears to embrace positions as disparate as reductive naturalism and vitalism. However, he grasped the fundamental problems posed by the mechanistic worldview and attempted to work out a viable alternative. I argue that his work does contain the resources for working out a nonreductive naturalism. I focus on his view of living being, his incorporation of evolution, and his value theory. In order to elaborate this view of nature, I draw on the work of subsequent thinkers such as Hans Jonas and Evan Thompson, both of whom seek an integration of phenomenology and biology and aim to restore the ontological autonomy of life. This kind of naturalism is "life-affirming" in the sense that it recognizes life as an autonomous kind of being irreducible to physiochemical properties and mechanistic causality. It holds that humanity is continuous with animal life and subject to evolutionary forces, yet resists the mechanistic, materialist interpretation of evolution. And, finally, it rejects the value-free vision of nature found in modern science, holding that all living things value in some sense. Heidegger, too, rejected the picture of nature bequeathed by modern science. However, he did not develop an evolutionary theory or anchor value in biology as Nietzsche attempted to do.

In the final chapter, I explain how the nonreductive naturalism sketched by Heidegger and Nietzsche might provide a conceptual foundation for an environmental ethic and a naturalized phenomenology. Most environmental thinkers would probably call themselves "naturalists"—but which kind of naturalism do they have in mind? If environmental ethics bases its view of nature on the natural sciences—scientific naturalism—it is arguably seeking for values in a valueless world, and is plagued by the problem of nihilism. A nonreductive naturalism solves this problem by anchoring value in the natural world through a phenomenological account of the organism's relation to its environment. Likewise, most contemporary phenomenologists would probably call themselves "realists"—but which kind of realism do they have in mind? I reject the agnostic posture toward metaphysics often struck by phenomenologists and hold that a nonreductive naturalism maintains a place for intentionality in the natural world. The vision intimated by Heidegger and Nietzsche reconstructs traditional views of nature as a great chain of being or *scala natura*, but does so without speculative supports and in a way that is consistent with evolutionary biology. I conclude by suggesting that this view of nature can support a "hierarchical biocentrism," which recognizes the intrinsic value of all living things while maintaining that higher, more complex forms of life embody greater value.

1

The Traditional Reading of Heidegger's Relevance for Environmental Philosophy and Ethics

Before we can address what our obligations to the environment might be, we have to determine what "the environment" or, more broadly, "nature," *is*. In other words, we have to determine what is traditionally called the "metaphysical ground of ethics" with regard to the natural world. While Heidegger was mainly concerned with the meaning of being, Michel Haar observes that "beginning with the Turn of the 1930s, both in *Introduction to Metaphysics* and *Origin of the Work of Art*, a new thought of elementary nature emerges under the names of *physis* and earth. This nature . . . turns out to be very close to being itself."[1] Though Heidegger's retrievals of the Pre-Socratic and Aristotelian accounts of *physis* only come to the fore in his work in and after the 1930s, it is a mistake to frame his philosophical interest in nature as merely a later development. Already in the early 1920s, Heidegger was mining Aristotle's works in hopes of finding and forging a model that more adequately describes human existence than the primal Christianity of Paul and Luther that had dominated his thinking up until that time.[2] The notion of factical life that Heidegger employs in some of these early lectures and that would serve as the backbone of his existential analytic in *Being and Time* is largely a phenomenological reinterpretation of some of the seminal concepts in Aristotle's *De Anima* and *Physics*, especially the concepts of life and nature. It turns out that for early *and* later Heidegger the meaning of being has much, if not everything, to do with the meaning of nature.

In this chapter, I first survey the numerous attempts to assess Heidegger's philosophy of nature and frame him as a protoecological thinker. Second, I analyze three unique aspects of Heidegger's approach to nature: his positions on anthropocentrism, axiology, and scientific naturalism.

Third, I describe the project of naturalizing phenomenology and explain the difficulties of naturalizing Heidegger's thought. I show that there are two phenomenological traditions—"transcendental" and "biological"—in tension in Heidegger's earlier work: the former is antinaturalist, while the latter points to a nonreductive naturalism. While he flirted with the second path, Heidegger ultimately chose the first. I submit that the traditional approaches to Heideggerian environmental thought—and Heidegger's later philosophy of nature—embrace the antinaturalist stance and face the four problems I mentioned in the preface: 1) Heidegger fails to develop a robust philosophical biology and deal with the "mind/life" problem; 2) he ignores evolutionary theory; 3) he rejects value theory; and 4) he regards nature as ultimately unintelligible.

Before turning to the history of Heidegger and environmental thought, let me state my positions on two issues that go to the heart of my critique of his approach: metaphysics and ethics. Regarding metaphysics, one might object that my attempt to "naturalize" Heidegger's thought and demand a philosophical biology from it is, to use his own phrase, a "round square and a misunderstanding." Indeed, it is not an understatement to say that Heidegger's chief concern was to *overcome* metaphysics. Would not a "naturalized" Heidegger entail some form of naturalism, that is, a metaphysical position to which his thought is fundamentally opposed?

"Metaphysics" is said in many ways; the Western tradition of substance metaphysics is not the only game in town. Two alternatives stand out in my mind: one, process philosophy, and two, forms of East Asian thought.[3] Regarding the first, we can consider the tradition of process philosophy[4] as an alternative metaphysical tradition, a minority report in Western thought that seems less opposed to Heidegger's thought than substance metaphysics. Indeed, I think Nietzsche's critique of the concept of being and embrace of the concept of becoming is one reason Heidegger sees great promise in his approach. But I see this as *differently* metaphysical, rather than *non*-metaphysical. Just so, Heidegger's eventual position regarding nature carries, I will argue, certain metaphysical assumptions, some of which are inherited from Kant. Put differently, one of my assumptions, which I hope to render plausible throughout, is that metaphysics is in some sense unavoidable.

I want to connect this with my critique of some of the so-called "postmodern" approaches to life, animality, and nature discussed toward the end of the book: just as we should be wary of what Derrida called the "metaphysics of presence," so should we avoid the "metaphysics of absence," of reifying the nothing as something "in itself." This is one reason, as I explain throughout, that Heidegger is wrong to dissociate science and philosophy.[5] Like Nietzsche, Whitehead incorporated the concept of evolution into his metaphysics; he recognized that this was a crucial insight furnished by science that had to be taken up and refashioned by philosophy and that this refashioning would in turn transform metaphysics. As such, Whitehead's philosophy of organism is not just a philosophical biology: it is also a metaphysics. And as I attempt to explain in the later chapters on Nietzsche, for him the "organic" never emerges—it is present all along. This is because Nietzsche rejects the mechanistic view of the world, including the mechanistic view of physics. Here, too, is where he departs from Heidegger. Like Kant, Heidegger took as normative a view of science locked in place in the seventeenth century, a view that set the course for and skewed modern biology. Rather than take a more expansive and quasi-Aristotelian view of science—which he flirts with in the 1920s—he dissociates it from philosophy proper, and this is in large part what causes him problems dealing with the concept of life: it renders impossible any kind of "continuum" view of nature and generates various forms of dualism.

My own view is that Heidegger tossed out the baby of metaphysics with the bathwater of substance metaphysics. He was led to do so in large part because of the rise of scientific naturalism in the modern period and its (in his view) roots in the origins of Western thinking. But the foundations of naturalism are problematic, and the current project of naturalizing phenomenology may open the way to an alternative naturalism. On one flank, it challenges the metaphysical neutrality of phenomenology by questioning how intentionality is anchored in organisms and emerged in evolutionary history, and on the other, it challenges the neo-Darwinian orthodoxy. I discuss this approach in the third section of this chapter, and draw heavily from one of its chief exponents, Evan Thompson, in the seventh chapter. All told, my contention is that the project of naturalizing Heidegger leads us to push against the postmodern pox on metaphysics and that the way forward lies in reckoning with the concept of life and the philosophy of biology.

A word is also in order about Heidegger's stance on ethics. Heidegger rejected the notion of supplying a theoretical ethics or of deriving an ethics from a metaphysics.[6] In this sense, the notion of deriving an environmental ethic from Heidegger's thought is potentially misguided. However, I think he goes astray in deflating ethics. Of course it is an option in theoretical space to reject the notion of an ethics, but it does not seem to me a good or well-founded one, nor one that squares with Heidegger's own position. For, as we will see, Heidegger *does* offer an ethics: we should let beings be, letting them unfold their own natural capacities, whatever those may be; our posture should be one of openness to the manifestation of being;

we should strive for a relationship with technology that does not corrupt nature or our humanity; and so on. Moreover, this ethics issues from his view of what the being-process *is* and *does*. These notions are not expressly formulated as ethical principles, but they do suggest an ethical orientation. One might cast this as a kind of Aristotelian prudence: the person who is properly attuned to the situation will simply respond appropriately. But this raises difficult questions: should we always let beings be, in all instances? What if there are conflicts between beings? How do we decide? When we apply these questions to environmental issues, I think the inadequacy of Heidegger's approach becomes clear.

And this is where philosophical biology and metaphysics become relevant. Heidegger correctly rejects the anthropocentric or, as I call it, the "projectionist" view of value; but he errs in jettisoning value as such. By saying we should let beings be and refusing to supply any criteria by which we can judge how to act in different situations, Heidegger seems implicitly committed to the notion that everything has equal value. This is one reason, I think, that he has been compared to deep ecologists, who embrace bioegalitarianism and face its attendant problems of differentiating higher and lower degrees of value. But if we anchor value in the organismic and ecological conditions of beings—as we find, for instance, in thinkers like Holmes Rolston, David Ray Griffin, and, I argue in later chapters, Nietzsche—then we have a way to distinguish higher and lower forms of value. I elaborate on this view in the final chapter.

Throughout, I contend that there is an overlooked strain of naturalism in Heidegger's thought, running from his early work on Aristotle to his engagement with the biologist Jakob von Uexküll and, finally, to his confrontation with Nietzsche. These untapped resources suggest a *naturalized* Heidegger that provides a sounder theoretical basis for environmental ethics than the traditional approach to his thought. But first, we need to review the history of Heidegger and environmental philosophy.

I. Heidegger and Environmental Philosophy: A Checkered History

The literature on Heidegger and environmental philosophy can be roughly divided into three groups: 1) early critics of the view of nature implied by his early analysis of human existence; 2) attempts to frame him as a deep ecologist or ecological thinker; and 3) continental studies and appropriations of his approach to nature, sometimes called ecophenomenology.

1. Early Critics of Heidegger's Account of Nature

Two of Heidegger's students, Hans Jonas and Karl Löwith, criticized him in 1966 for being an existentialist with an anthropocentric understanding of nature. Jonas, attempting a phenomenology of life, charged that the early Heidegger espoused a Gnosticism in which humans were ontologically dissociated from nonhuman beings and nature as a whole, and was hence a prisoner of the very Cartesian dualism that he was trying to overcome. Jonas's outlook on the possibility of a Heideggerean natural philosophy was unequivocal: "No philosophy of nature can issue from Heidegger's thought."[7] In a similar vein, Löwith held that "the criticism of the Cartesian ontology [in *Being and Time*] rests also on the distinction of two kinds of being which are different in principle: human *Dasein* and entities."[8] He also claimed that Heidegger's existentialist notion of history betrayed his enslavement to the modern scientific "mathematization" of nature in which human beings have no proper place. Once the notions of nature as cosmos and creation fell away, objective, value- and logos-free nature was all that remained, and historicism and existentialism came into being. As Löwith puts it, "if the universe is neither eternal and divine (Aristotle) nor contingent but created (Augustine), if man has no definite place in the hierarchy of an eternal or created cosmos, then, and only then, does man begin to 'exist,' ecstatically and historically."[9] The result, he claims, is that nature is deemed beyond the pale of legitimate philosophical inquiry in *Being and Time*. Löwith's capital conviction is that the inadequacy of Heidegger's account of nature lies in his understanding of history, and this because the latter is approached hermeneutically as a horizon of sense that conditions everything humans encounter, including nature itself.

Both of these early critics of Heidegger's approach to nature perceive the need for a return to (Löwith) or a revision of (Jonas) something like the traditional great chain of being and the notion of nature as a cosmos, and both fault him for being too anthropocentric. Since virtually all of the more recent attempts to wed Heidegger and environmental thought see him as a nonanthropocentrist, and since many of these saw the union of the two camps as relatively unproblematic, it is imperative to keep the concerns of these early critics in mind. Indeed, I hope to show that Jonas and Löwith were basically on the right track: Jonas saw that Heidegger did not adequately grapple with the ontological status of life, while Löwith saw the problems that Heidegger's view of history and rejection of a *scala natura* caused for his philosophy of nature. In chapter 7, I provide a protracted overview of Jonas's view in tandem with that of Evan Thompson; in my view, these thinkers develop with greater sophistication the naturalism sketched by Heidegger and Nietzsche.

2. Heidegger and Deep Ecology

The second wave in Heideggerian environmental thought involves the attempt to establish a connection between Heidegger and deep ecology, the school that has been most often compared with Heidegger's thinking about nature.[10] Deep ecology is a broad term canvassing both an intellectual movement and a political cause that can be loosely defined as a group of individuals committed to the notion that the status quo in the relationship between humanity and nature is detrimental to both and that only a radical reorganization of society can bring about the needed change. In environmental philosophy, the chief representatives of deep ecology are the late Norwegian philosopher Arne Naess (who coined the phrase and more or less founded the movement), American thinkers Bill Devall and George Sessions, and Australian philosopher Warwick Fox. Though there is some dispute over the essentials of deep ecology,[11] Naess insists that the personal and pluralistic nature of the movement, that is, its ability to accommodate and incorporate inspiration from different cultural, religious, and intellectual perspectives, is one of its strengths. There appears to be a consensus that espousal of the eight-point "Deep Ecology Platform" enumerated in 1984 by Naess and Sessions is a necessary condition for calling oneself a deep ecologist. This platform states that human and nonhuman life, as well as human and nonhuman collectives, including species, natural habitats, and human cultures, possess inherent worth and that biodiversity is an intrinsic value. It is also committed to the ideal of "bioequality," the belief that all living things have equal moral worth.

Deep ecologists are convinced that the adoption of their platform is impossible without a radical, deep shift in humanity's self-identity: their goal is a self-identification with nature. This emphasis on a radical transformation in human subjectivity is what makes deep ecology a good candidate for comparison with Heidegger, who was likewise convinced that an ontological shift—a drastic change in humanity's understanding of being—is required for humans to appropriate their past and live authentically (the early Heidegger) and dwell properly on the earth and stem the erosive tide of modern technology (the later Heidegger). Like deep ecologists, Heidegger's environmental philosophy is not centered on criticizing traditional Western moral philosophy and furnishing a new theoretical ethics that includes nonhuman beings, but on rethinking our understanding of nature as a whole by criticizing traditional Western metaphysics.

If for Aristotle metaphysics is first philosophy, and if for Levinas ethics is first philosophy, perhaps we may say that for Heidegger "physics" is first philosophy: physics not in the sense of modern natural science and materialism, but in the sense that its root, *physis*, had for the Greeks. Heidegger maintained that for the Greeks "*physis* is being itself" and it

> originally encompassed heaven as well as earth, the stone as well as the plant, the animal as well as the man, and it encompassed human history as a work of men and the gods. . . . *Physis* means the power that emerges and the enduring realm under its sway. . . . *Physis* is the process of arising, or emerging from the hidden, whereby the hidden is first made to stand.[12]

As I will explain later, this identification of being with *physis*, a move that comes to define Heidegger's later work, is the mainspring of environmental—and especially deep ecological—interpretations of his work.

In the early 1980s, Michael E. Zimmerman argued that Heidegger's critique of modern technology's reduction of nature to raw material or "standing reserve" (*Bestand*) purely for human purposes and his notion of "letting be" (*Gelassenheit*) offered a way out of the domineering and exploitative attitudes and practices responsible for the ecological crisis.[13] In claiming Heidegger to be a biocentrist, deep ecologists seized upon fixtures in his later philosophy, such as the elevation of poetic, meditative thinking over rational, calculative thinking, his affection for the Pre-Socratics, his critique of the enframing (*Gestell*) of modern technology, and his call for humans to learn how to dwell poetically in the fourfold of earth, sky, gods, and mortals.[14]

These thinkers looked to Heidegger's later philosophy rather than *Being and Time* or his earlier work because of his alleged "turn" after the 1920s. Deep ecologists' and Zimmerman's early work on Heidegger tend to assume a facile distinction between an early, anthropocentric Heidegger (subject to Jonas' and Loewith's critiques) and a later, nonanthropocentric Heidegger (immune from those critiques). While useful heuristically, this distinction cannot be so easily made, since many of the mainsprings of Heidegger's later thought that environmental philosophers tend to seize upon were already nascent in his early pre-*Being and Time* works, especially his concerns with Aristotle. Zimmerman asserts that Heidegger's "later phenomenology, ever more hermeneutical in orientation, amounted to a radical uncovering of insights gained by the phenomenological ontology of previous great thinkers, above all Aristotle. Heidegger interpreted crucial Aristotelian concepts, such as *physis, energeia, dynamis, kinesis,* and *metabole*."[15] This overlooks the fact that Heidegger was already mining Aristotle's works for these insights in the early twenties.[16] As such, the so-called "turn" can, as John Van Buren has put it, be seen as a "re-turn" to elements already laced within Heidegger's early formulations of the question of being, and these include a concern for a more poetic, nonanthropocentric sense of nature.[17]

Deep ecologists' and Zimmerman's optimistic outlook on a Heideggerian environmental philosophy and ethics was followed by a cluster of essays and books that were rather sanguine about the attempt to frame Heidegger as a deep ecologist or as a nonanthropocentric ecological thinker, grounding the project in his critique of humanism,[18] his reinterpretation of Aristotle,[19] his account of language,[20] or his unique understanding of dwelling on the earth.[21] The most extensive and important work in this camp is Bruce Foltz's 1995 book-length study of Heidegger's "metaphysics of nature." Foltz provides a meticulous analysis of Heidegger's early and later writings on nature, contests Jonas' and Löwith's charges of anthropocentrism, and argues that through his critique of the modern scientific view of nature and its roots in the Western metaphysical tradition's interpretation of nature, Heidegger unearths a different sense of nature that can underwrite an environmental ethic. Though Foltz does not attempt to paint Heidegger as a deep ecologist, he admits the affinities between them:

> All these approaches [i.e., ecocentrism and deep ecology] share with Heidegger a sharply critical orientation toward the "subjectivity," individualism, and humanism of modern consciousness; all see the need for radical change in life, thought, sensibility, or culture; all see human beings as properly understood only within the context of, and hence in some sense subordinate to, something greater.[22]

Despite these affinities, some scholars, including and especially Zimmerman himself, became suspicious of the compatibility of the two approaches.[23] These suspicions were bred in part by Heidegger's political entanglements, his espousal of a nonprogressive understanding of history, his rejection (or at least circumvention) of Darwinism, his reservations about science, his ambiguous interpretation of animals, and his insistence that humanity is ontologically separate from nature. Though he admits affinities between the two approaches, Foltz himself observes that one problem with deep ecology is that it "is so captivated by the scientific viewpoint that it deals with the task of learning to dwell *within* as something to be defined objectively from *without*, vis-à-vis an explanation of human behavior as properly functioning components of a healthy ecosystem."[24] Heidegger's approach is different in that he endeavors to rethink being and human being in terms of temporal structures and meaningful relations founded in human experience, not in terms of ecosystemic relations discovered *via* the natural sciences. Another way of saying this is that, while Heidegger does seek to situate human beings within a more holistic, relational ontology, he casts them as members belonging within a meaningful world, whereas deep ecology plucks humans off the top of the allegedly natural hierarchy only to insert them as a part of an objective whole, a node in a system of integrated functions. As Zimmerman points out, this is basically due to Heidegger's position that there is an ontological abyss between humans and nonhumans: "Heidegger's perceived anthropocentrism, his concerns that the [deep ecology platform] manifests modernity's control impulse, and the fact that some deep ecologists adhere to progressive views of history, indicate problems in attempts to read Heidegger as a forerunner of deep ecology."[25] Note that the concerns of the Heidegger-deep ecology critics are similar to the early critics of his account of nature. Both point to the need to critically transform Heidegger's thought if it is to contribute to environmental philosophy and ethics, or reject it if this is not possible.

3. Continental Approaches to Heidegger and Environmental Philosophy

The third major branch of environmental philosophy inspired and influenced by Heidegger comprises a set of books by continental scholars, most of which can be classed as early attempts at an ecophenomenology. This field, which was consolidated in 2003 with the publication of the anthology *Eco-Phenomomenology: Back to the Earth Itself*, attempts to apply the phenomenological approaches of Husserl, Heidegger, Merleau-Ponty, and others to questions of environmental theory and practice.[26] I will explain the project of ecophenomenology in more detail below, but here I want to briefly summarize some of the early and noteworthy works in the field and mark their use of Heidegger.

In *The Embers and the Stars* (1984), Erazim Kohak sought to rehabilitate the "moral sense of nature" by drawing on Husserl's and Heidegger's phenomenologies in order to deconstruct the dissociations of nature/culture and fact/value. For Kohak, "*Prima philosophia* cannot start with speculation. It must first see clearly and articulate faithfully the sense evidently given in experience."[27] By cultivating or perhaps rekindling a breadth and depth of vision that allows natural beings to show themselves in their fullness, rather than just as objects for investigation by science or manipulation by technology, Kohak thinks we can prepare the ground for an environmental ethic.

John Llewelyn (1991) and David Abram (1998) both employ Heidegger in order to recapture the alterity of the natural world.[28] Llewelyn attempts to enlist Heidegger and Levinas together in the cause of cultivating a poetic "middle voice" that is neither wholly active toward and constitutive of the nonhuman nor entirely passive and subject to it. As he explains, "Given the ecological interdependence of things, human and nonhuman, other non-human beings no less than other human beings have a claim upon me through their simply being needy beings other than me. . . . [T]he naked alterity of a finite vulnerable thing suffices to put me under a direct responsibility toward it."[29] Thus he plays Heidegger off against Levinas by extending the notion of the ethical relation to the other as constitutive of the subject to the nonhuman, natural order, and plays Levinas off against Heidegger in order to supplement the latter's ontological focus with an ethical orientation based on need. Abram, drawing on Husserl's notion of the earth as humanity's "primitive home," Merleau-Ponty's investigations into the body subject and the reciprocity between self and world, and Heidegger's views on space and time, claims that a "return to our senses"—to the depth and complexity of our immediate sensual experience of the natural world—can recover a relationship to animate nature and rupture the one-sided anthropocentrism that restricts meaning to the human realm.

Two important works that deal less with Heidegger's views on nature in general and more with specific ecological themes are David Farrell Krell's *Daimon Life: Heidegger and Life Philosophy* (1992), which examines Heidegger's understanding of life, and Brett Buchanan's *Onto-Ethologies: The Animal Environments of Uexküll, Heidegger, Merleau-Ponty, and Deleuze* (2008), which examines Heidegger's appropriation and critique of the biology of his time, especially conceptions of the organism and the *Umwelt* ("environment" or "surrounding world").[30] Krell submits that the phenomenon of life permeates the dominant themes in Heidegger's early and later thought and that "however much Heidegger inveighs against life-philosophy his own fundamental ontology and poetics of being thrust him back onto *Lebensphilosophie* again and again."[31] The being of life haunts Heidegger's thought. His most sustained engagement with theoretical biology, animals, and life is found in his 1929–30 lectures, *Fundamental Concepts of Metaphysics*. In my view, Heidegger was on the right track in attempting to work out a nonreductive view of life that comprises the human and nonhuman; after these lectures, however, he abandons biological questions and takes a different path. Buchanan likewise shows the considerable effect that biological considerations, particularly Jakob von Uexküll's inquires into the animal *Umwelt*, had on Heidegger's notion of being-in-the-world and investigations of animal being. Like the being of life, the being of animals and the environment bedevils Heidegger's work, and he never quite reconciles them with human being.

II. Heidegger and Environmental Philosophy: A Round Square?

Before delving into the texts themselves, I want to sketch a few more of the broad contours of Heidegger's approach to nature. While I am here referring mainly to *Being and Time* and trends in Heidegger's thought leading up to that work, many of these themes echo in his later works, and I will indicate where this is the case. To the extent that Heidegger concerns himself with nature in *Being and Time*, he does so not in order to formulate an environmental ethic and assign orders of value to nonhuman creatures, but rather in order to disable unfounded senses of nature, to allow all phenomena—whether we conceive of them as natural, as cultural, as spiritual, as artifactual, etc.—to arise just as they are and to describe them as such.

As I indicated above, Heidegger's approach to nature is unique. Zimmerman points out that his views differ from those of mainstream Anglo-American environmental philosophers in at least three basic and linked ways: 1) he is neither a (conventional) anthropocentrist nor a biocentrist, 2) his approach is not axiological, but ontological, and 3) he does not take the worldview of scientific naturalism for granted. I examine these one at a time.

1. Anthropocentrism

The terms of the debate over which beings possess inherent or instrumental value is mainly waged between anthropocentrists and biocentrists. As I mentioned above, the early Heidegger was criticized as an anthropocentrist by Jonas and Löwith for treating nature as a correlate of human consciousness, that is, as something constituted by human intentionality that only shows up and has sense within the horizon of an historical human world. They held that Heidegger reduced nature to a field for human projects and thus gave tacit assent to the reductive, materialist view of nature in modern natural science. But as Bruce Foltz points out, Heidegger's "interpretation of nature as *Vorhandenheit* [presence-at-hand], and his critique of the concept of nature as obscuring our understanding of both ourselves and the world, are in fact an interpretation and a critique of the metaphysical concept of nature rather than a disparagement of the phenomenon itself."[32] In other words, Heidegger's phenomenology in *Being and Time* brackets or suspends ontological claims about nature in order to allow a more original encounter with nature to emerge.

This is just a specific application of his general method throughout the book. Heidegger's qualm with anthropocentrism is not primarily that it prioritizes human beings over nonhuman, natural beings, or that it holds that humans are the sole source and bearers of intrinsic value, but that it ignores being itself. He thus criticizes anthropocentrism, but for different reasons than most environmental ethicists: his interest is ontological, not ethical. To identify what is unique about his approach to nature, let us look at his starting point (human historical *Dasein*), his method (hermeneutic phenomenology), and his position on axiology, which I explore below.

2. Axiology

The second distinguishing mark of Heidegger's approach to nature, his aversion to axiology, pertains to both his early and later work. His early inquiries into the nature of value were centered on a critique of the views on logic, truth, value, and judgment of Rudolf Lotze and neo-Kantians Heinrich Rickert and Wilhelm Windelband. Parvis Emad has shown how these early inquiries prefigure Heidegger's critique of the tendency in Western metaphysics, which culminates in Nietzsche, to think being as constant presence: "[Heidegger's early] Lectures clarify the position of *Being and Time* on the ontological status of value: when taken as a mode of affirmation, the being of value is conceived as constantly present in a valid proposition."[33] Heidegger's strategy here is the same as that used with the concept of nature above: to question the ontological status of a concept, in this case value, in order to bring to light the unexamined prejudices that motivate the claims surrounding it. Thomas Nenon enumerates three reasons why Heidegger rejects value theory:

> To put it in "isms": on the one hand, the theory of values as developed in Neo-Kantianism is propositionalist (because of its orientation upon judgment), representationalist (because of the primacy of the theoretical judgment both as the fundamental building block of mental activity and also as the form of judg-

ment in which philosophy realizes itself), and intentionalist (because of its emphasis upon consciousness' ability to be present to itself in reflection).[34]

Heidegger's answers to these "isms" are, respectively: 1) He shifts the focus of analysis away from propositions and back to pretheoretical involvement in a world of shared meanings and practical involvements, that is, to what he famously calls "being-in-the-world." 2) He argues that, *contra* the modern epistemological tradition issuing from Descartes, we should not see consciousness as a self-enclosed container, a subject that represents the world through images and concepts and then re-presents it through statements in speech or writing, nor should we see truth as the correct correspondence between representations and reality. Instead, we should see consciousness as always already entangled with the world, as always *our* consciousness, and as disclosing, enacting, and bringing forth that world. This leads to 3) Heidegger's critical appropriation of Husserl's phenomenology, that is, his turn to hermeneutics. We cannot ever conceptually grasp the contents of our own mind or the world through reflection because the latter stance is founded on and made possible by our prior involvement in a world. This is a world into which we are thrown, a horizon of meaning whose other side we cannot access because it is the means by which we access anything at all. Moreover, since consciousness is always already out there in the world, it can never "catch up with" itself and always "runs ahead of" itself. It is intrinsically self-transcending, and thus existence cannot be "paused" in order to objectify it and provide a full catalogue of its structure. So Heidegger's qualm with value theory is ultimately ontological: it rests upon what he takes to be the misguided tendency to interpret being as constant presence, to basically look at only one-half of things, and to force them to conform to the way they appear to us.

To relate this general position on values to the concept of nature, in *Being and Time* Heidegger says, "In interpreting, we do not, so to speak, throw a 'signification' over some naked thing which is present-at-hand, we do not stick a value on it; but when something within-the-world is encountered as such, the thing in question already has an involvement which is disclosed in our understanding of the world."[35] He does not accept the "fact-value distinction," according to which nature is the realm of value- and meaning-neutral, objective facts, and subjectivity, psychology, or consciousness is the realm of values, which humans posit or project *onto* mere things. This position is a cognitive achievement, not a self-evident given.

Nature must first be "set up" and "framed" as an objective order; we do not actually encounter it as such. Nor does he accept the position that values "really" inhere in things as qualities or properties. Stripping nature of values and stuffing it full of them stem from the same mistake: failing to see that our access to nature depends on our prior, prereflective involvement in a *world*—a world that we primarily encounter as neither merely cultural nor purely natural. Speaking of the later Heidegger, Zimmerman elaborates: "[Heidegger] maintains that the very concept of 'value' arose along with the power-hungry modern subject. Hence, extending value to non-human beings encompasses them within the same subjectivity that is central to technological modernity."[36] All of these issues—value thinking, subjectivism, humanism, anthropocentrism, modern technology, being as presence—are tightly constellated in Heidegger's later work, and I explore them at length in chapter 5. For now, suffice it to say that Heidegger's opposition to value thinking places him at odds with the better part of environmental philosophy. In later chapters, I suggest that this opposition is one of the main problems in his philosophy of nature.

Given that one of Heidegger's main problems with value thinking is that it passes over and neutralizes what he terms the phenomenon of world and that the motive for a phenomenology is to recover the original, founding experiences that give birth to and underlie our working concepts of nature, we can now look at what makes Heidegger's approach to nature phenomenological and, more generally, detail just what distinguishes a phenomenological approach to nature from other approaches. But first, let us see how this bears upon the third distinguishing mark of his approach: the rejection of scientific naturalism.

3. Scientific Naturalism

The term "naturalism" can mean many things. It could mean "methodological" naturalism, which eschews metaphysical claims and merely purports to study "empirical" phenomena, things that can be observed with the senses or the instrumental extension thereof. It could mean "metaphysical" naturalism, which makes the stronger claim that only empirical phenomena are real. According to Keith Campbell, "metaphysical naturalism affirms that the natural world is the only real one, and that the human race is not separate from it, but belongs to it as a part. . . . The natural world is the world of space, time, matter, energy, and causality."[37] In a recent anthology devoted to the theme of naturalism, Mario De Caro and David Macarthur label this view "scientific naturalism" and develop it in detail: "[S]cientific naturalists typically conceive nature as a causally closed spatio-temporal structure governed by efficient causal laws—where causes are thought of, paradigmatically, as mind-independent bringers-about of change or difference."[38] Despite the many varieties of scientific naturalism, they suggest that there are two main themes. One is ontological—"a commitment to an exclusively scientific conception of nature"—while the other is methodological—"a reconception of the traditional relation between philosophy and science according to which philosophical inquiry is conceived as continuous with science."[39] As the authors point out, scientific naturalists not only "claim that the conception of nature of the natural sciences is very likely to be true," but go further by insisting that "this is our only bona fide or unproblematic conception of nature."[40] In this way, scientific naturalism is one of the most entrenched default settings in contemporary philosophical discourse.[41]

Naturalism could also mean "romantic" naturalism of the John Muir variety, which endorses something like an original kinship between human beings and the natural world, waxes poetic about natural landscapes, and sometimes bears ill will toward modern technology and industry. Many environmental philosophies can rightly pass as "naturalist" in this third sense.

For the sake of clarity, when I use the terms "naturalism" or 'naturalist," I am usually using it in the second sense of metaphysical or scientific naturalism.[42] The important connotations of the terms for the analyses to follow are the ideas that humans are just another animal species, that neo-Darwinian evolution operating through random mutation and natural selection is basically correct, and that we cannot maintain a teleological view of life or nature. So what I have in mind here is more like "biological naturalism" or "biologism." This is different from materialism or physicalism. While Heidegger was concerned to avoid collapsing the region of life studied by biology and ecology to physics and chemistry, he was just as intent on preserving the autonomy of sense, logic, and intentionality from reduction to psychology. So Heidegger would disagree with Campbell's claims that "[i]t is possible to affirm naturalism while insisting that the higher faculties in humans and other animals cannot be given a physicalistic reduction, and nonmaterialistic naturalism avoids the difficulties that materialism has, for example, in accounting for the intensional characteristics, such as linguistic meaning and psychological understanding."[43] He would disagree because, to him, it does not matter whether you restrict reality to what can be accessed by physics or biology or psychology—you are still defining reality according to the category of actuality from the standpoint of a theoretical attitude that overlooks its prior, pretheroretical involvement in a meaningful world. Any naturalism that stakes its claims on the deliverances of the natural sciences operates on the order of explanation, restricting reality to that which is governed by causality. Changing the causes from physical forces to biological instincts or environmental pressures does not do away with the underlying problem: accounting for meaning and intentionality. The regions, the contexts in which particular beings are studied in the various sciences, are regions of *sense*, so the search for explanations presupposes that beings are *intended* in a certain way by the researcher. Heidegger is consistently adamant that human existence cannot be adequately conceived of in biological categories such as instincts or drives. Biologism may be less reductive than physicalism, but for Heidegger it fails to heed the ontological difference between being and beings as well as the ontological gulf between humans and animals. And the recognition of the ontological difference, for Heidegger, lies in the phenomenological reduction. With Heidegger's opposition to naturalism laid out, we can now turn to the question of whether his thought can be naturalized in any way.

III. Eco-phenomenology: A Naturalized Heidegger?

Perhaps the central feature of a phenomenological approach to nature is a staunch opposition to scientific naturalism. As Ted Toadvine explains in an anthology devoted to eco-phenomenology,

> One point of agreement among phenomenologists is their criticism and rejection of the tendency of scientific naturalism to forget its own roots in experience. The consequence of this forgetting is that our experienced reality is supplanted by an abstract model of reality. . . . The return to "things themselves" and the critique of scientific naturalism both point in the direction of much contemporary environmental thought.[44]

Put differently, the naturalist tends to reduce the data of experience to data *as defined by* a scientific discipline, be it physics, biology, or ecology. She tends to take for granted that the phenomena she is investigating are real, exist independently of the mind, and are perhaps even the only things that are real. First- and second-person experiences are thus explained in third-person terms.[45] Thus, for the naturalist, the ways in which humans usually experience and interface with nature and natural beings fall outside the scope of legitimate inquiry, since they are merely "subjective." The naturalist is obliged to regard any meanings or values that humans claim to inhere in natural phenomena as nothing more than the expression of psychological or cultural attitudes that tell us nothing about nature as it is "in itself."

To be fair, the naturalist does this with good intentions: she is aiming to bracket whatever psychological and cultural beliefs, dispositions, and prejudices may skew her perception and cloud her judgment in hopes of arriving at objective truths about her subject. The phenomenologist, however, points out that what the naturalist takes to be a "view from nowhere" is always, in truth, a "view from somewhere." As Lester Embree, applying Husserl's phenomenology to nature, explains, the first task of a constitutional phenomenology of the environment is "to provide the analysis in terms of which the 'nature' correlative to the naturalistic attitude is an abstract part of the cultural world."[46] For the phenomenologist, he continues, "the environment is first of all part of the cultural world—that is, made up of objects that not only have a naturalistic foundation that is vital or organic, but are also valued and willed in pre-theoretical human life."[47] Nature or the environment, then, is not something we are "in," in the sense that a table is "in" a kitchen; rather, nature is partly constituted by and inconceivable without a knowing subject and a community of knowing subjects. What the naturalist touts as an objective nature bereft of human value positings, be they moral or aesthetic, turns out to be—at least in part—a construct, a correlate of consciousness that is, in truth, derivative of and founded on a more basic mode of experience that Husserl calls the "life-world" and that Heidegger in *Being and Time* calls "being-in-the-world." The "theoretical-naturalistic attitude" is rooted in the "natural attitude," and the latter is not taken as the antipode of "culture." What we separate and oppose as culture and nature are actually just different aspects of an experiential totality, and an eco-phenomenology aims to clarify the structural relationship between them. It should be clear, then, that phenomenology is germane to environmental philosophy, since it aims to lead our attention back to—and to reinstate—a meaning- and value-laden experience of nature and natural beings that is prior to the "commonsense" dissociation of culture and nature.

As it has been loosely defined thus far, the project of an eco-phenomenology seems one that Heidegger would likely endorse. But in order to set the stage for his iteration of phenomenology, let us take a closer look at some of the specifics of eco-phenomenology as put forth by Toadvine. Toadvine states that eco-phenomenology rests on two claims: "[F]irst, that an adequate account of our ecological situation requires the methods and insights of phenomenology; and second, that phenomenology, led by its own momentum, becomes a philosophical ecology, that is, a study of the interrelationship between organism and world in its metaphysical and axiological dimensions."[48] Heidegger would likely not quibble with the first claim, but he would definitely take issue with the second, if "philosophical ecology" is so defined, and for two reasons. First, as I discussed above, Heidegger never wavered in his opposition to axiology of any kind. In *Being and Time* he inveighs repeatedly against the Neo-Kantian value theory prevalent in his own day,[49] and in the "Letter on Humanism" he proclaims that "thinking in values is the greatest blasphemy imaginable against Being."[50] For Heidegger, axiology is a symptom of anthropocentric humanism—a stance he later comes to consider all but synonymous with metaphysics—which discloses beings in a one-dimensional way as fodder for human purposes and interests. This is also the major motor of his rejection of Nietzsche's philosophy and casting of the latter as the last metaphysician at the nadir of nihilism.

The second reason Heidegger might have misgivings about this project has to do with the ontological status of nonhuman beings, and this concerns a rift within the phenomenological tradition itself. Marjorie Grene has argued that there are actually two phenomenological lineages. The first, which has dominated continental philosophy in the U.S. and Europe, is "transcendental phenomenology," introduced by Husserl and modified by Heidegger, Sartre, and others. This approach is characterized as transcendental because, following in the modern tradition of Descartes and Kant, it tries to determine the a priori universal and necessary structures of human knowledge and existence, and thus the parameters of knowledge. The point of departure here is always the experiencing subject. Whether the latter is conceived of as a transcendental subjectivity (Husserl) or as *Dasein* (Heidegger), the general tendency in the transcendental approach is to posit an ontological separation between the human and the non-human, the order of consciousness and the order of nature. Thus, it is argued, the tradition of transcendental phenomenology retains and reformulates Descartes' ontological prejudice about the relationship of mind and world. While it may prove effective in criticizing naturalism, it has difficulty integrating human beings with the animal, the organism, the biological, the living, and so forth. Indeed, one could see the existentialist leanings of Heidegger and Sartre, who emphasize the singularity, if not oddity, of humanity's place in nature—as manifested by our anxiety, radical freedom, and sense of not being at home in the world—as a symptom of this difficulty; it is exactly this strain in Heidegger's thought at which Jonas's and Löwith's critiques are aimed.

The second and less well known phenomenological tradition is what we might call "biophenomenology" and comprises a group of thinkers that includes Helmut Plessner, Max Scheler, Jakob von Uexküll, Jonas, Marjorie Grene, and Neil Evernden.[51] These thinkers sought to investigate the theoretical underpinnings of biology and drew on cutting-edge developments in the field that suggested a break from scientific naturalism and a picture of nature more akin to and perhaps compatible with the understanding of consciousness being advanced by phenomenology. They also sometimes involved retrievals of premodern traditions; Jonas, for instance, set himself the task of rewriting Aristotle's *De Anima* postevolutionary theory. The basic goal of these thinkers seems to have been the establishment of a "biology of subjects" in which the phenomenological category of intentionality—the structural correlation of consciousness and world—is extended beyond human beings and down to animals, plants, and life as such. Thus, the organism is approached not as a machine or as an objective system either merely motored by the commands programmed into its genetic structure or purely reacting to stimuli in its environment, but rather as a subjective being with intentionality that in part brings forth and co-constitutes and even values its environment. The goal here is to extend some form of consciousness, intentionality, or interiority, no matter how primitive, "all the way down" and thus situate human beings along a common continuum with nonhumans. This is what Toadvine means when he describes "a philosophical ecology, that is, a study of the interrelationship between organism and world in its metaphysical and axiological dimensions."[52] Ecophenomenological naturalism is thus carrying on this tradition.

There is clearly an Aristotelian strain here: the aim to recapture a teleological view of nature. However, these thinkers were also by and large convinced that the basic insights of phenomenology must somehow be squared with the theory of evolution. Since Aristotle's nature is eternal, his natural kinds (arguably) fixed and unchanging, and his outlook on nature clearly nonevolutionary, a mere revival of Aristotle's philosophy of nature in opposition to modern mechanistic materialism is not sufficient. The insistence of the thinkers in this tradition on the need for philosophy to take evolution seriously and free it from a materialistic and naturalistic ontology is, as I discuss in future chapters, one of the major lacunae in Heidegger's approach to nature. This is ironic, however, because many of these thinkers were deeply influenced by Heidegger. Indeed, as I explain in the next chapter, in the early 1920s Heidegger sketched a Neo-Aristotelian ontology of life informed by Uexküll's anti-Darwinian biology and patterned on the *De Anima*, and he seemed to embrace something like the phenomenological naturalism that Toadvine has called for. Yet this project was soon scrapped, and Heidegger turned down the "transcendental" path. By "transcendental,"

I do not mean Kantian (since Heidegger explicitly repudiates Kant's transcendental philosophy by the end of the 1920s); rather, I mean his adoption of certain assumptions about nature that he inherits from Kant. Heidegger follows Kant in deeming nature "in itself" a mysterious flux, a blooming buzzing confusion that exceeds our categories of understanding (this is what he will call the "law of the earth"). The notion of nature connected with Kant's concept of the sublime, I argue, is very much like earth, *physis*, and fourfold, some of the later Heidegger's choice terms for nature. Heidegger agrees with Kant that science "allows nature to be heard" (for Kant, space and time are empirically real, but transcendentally ideal—physics is not just poetry), though it cannot exhaust the richness of nature, with science predominantly understood here as mathematical physics that discovers universal and necessary laws. In short, Heidegger does not quite know what to do with the category of life: his Aristotelian leanings push him toward a "metaphysical interpretation of life," but his Kantian leanings push him to be skeptical about the metaphysical pretensions of biology and conclude that no comparison between the human and the animal/living is possible and that there is an unbridgeable ontological gap between them. My point is that during the 1920s, Heidegger shifts from a Neo-Aristotelian realism to a kind of Kantian skepticism about whether certain biological categories are constitutive principles of nature, rather than just structures of human understanding. While it is true that he disavows Kantian philosophy and thinking from the perspective of "subjectivity" as a residual form of metaphysics (this critique later comes to comprise Nietzsche as well), the elements of Kant's thought stated above seem to persist in his thought. I discuss this in greater detail in the conclusion of the fourth chapter.

The transcendental approach to phenomenology effectively "puts the natural sciences in their place" by disabusing them of their ontological pretensions through showing that they are founded on structures of experience for which they cannot themselves account and thus draws a clear demarcation between philosophy and the nature of knowing and consciousness, on the one hand, and the natural sciences and their objects of study, on the other. The bio-approach, however, wants to know how these two realms fit together. This brings us to the early Heidegger's ambivalent relationship to this second phenomenological tradition. On the one hand, Heidegger at times devotes attention to issues surrounding biology, life, and the organism and draws, for example, on Uexküll's revolutionary notion of the *Umwelt* in order to criticize mechanistic approaches to biology. In 1929, he even goes so far as to say that animals in some sense have a world.[53] Indeed, throughout the early 1920s, this seems to be his dominant position. On the other hand, though, his overriding concern with the question of being involves what it means for humans and how it has evolved and been answered throughout human history. He tends to stress what separates humans from animals and nature rather than what unites them, and, as I mentioned above, despite early forays, he never quite comes clean about the ontological status of life. It is a theme that recurs in his various engagements with Aristotle, his forays into the metaphysical biology of Uexküll, and most dramatically, in his confrontation with Nietzsche.

Conclusion: Naturalizing Heidegger

While Heidegger's ambitious overhaul of the Western philosophical tradition is in many ways of a piece with some approaches to environmental philosophy, and though in places he appears to offer the fundaments of a nonanthropocentric ethic through notions such as "letting be" (*Gelassenheit*), his approach to nature is unique and can only be assimilated with or collapsed into extant approaches with great difficulty and distortion. His analysis of human existence in *Being and Time*, his "History of Being" explicated in works from the 1930s onward, and his ambiguous position on the ontological status of life, animals, and the latter's relation to human beings paint a dualistic picture of the relationship between humanity and nature.

In my analyses of nature-related themes in his early and later works, of his accounts of nihilism, and of his treatment of the concepts of life and the animal, I aim to expose the four problems with his philosophy of nature: 1) Heidegger fails to develop a robust philosophical biology and deal with the "mind/life" problem; 2) he ignores evolutionary theory; 3) he rejects value theory; and 4) he regards nature as ultimately unintelligible. While Heidegger does effectively critique scientific naturalism, he does not resolve the residual problem of articulating a vision of nature in which humans, animals, and other beings fit along a common ontological continuum. Though he offers promising sketches of a Neo-Aristotelian philosophy of life in his earlier work, appropriates the ideas of Uexkülll, and finds in Nietzsche's life-affirming naturalism a counterweight to modern nihilism, he abandons these projects and embraces an antinaturalist philosophy of nature that, while rhetorically inspiring, cannot provide a viable foundation for an environmental ethic. In the following chapters, I trace the development of his philosophy of nature and explain how the naturalistic resources in his thought can mitigate the problems mentioned above. My hope is to relieve the shape of a naturalized Heideggerian environmental philosophy.

2

The Question concerning Biology

*Life, Soul, and Nature in Heidegger's
Early Aristotle Lecture Courses*

In this chapter, I trace Heidegger's investigations into life and animality in the lecture courses on Aristotle in the early-mid 1920s. Here, I stress three factors: 1) the influence of biologist Jakob von Uexküll on Heidegger's reading of Aristotle; 2) Heidegger's attraction to a "continuum" view of the human-animal and human-nature relationship; and 3) how the Aristotle lectures prefigure *FCM*, his most sustained attempt at a philosophical biology.

Moreover, in this chapter I begin an argument that will be developed over the next four chapters. I submit that we can triangulate Heidegger's position on the ontological status of life by tracing the tension between the Kantian and Aristotelian strains in his work. On the one hand, Heidegger follows Kant in refraining from claiming teleology as a constitutive principle of living being and in eschewing a robust metaphysical biology; as I show in the next chapter, the Kantian strain begins to dominate in *Being and Time*. On the other hand, Heidegger sees Aristotle's understanding of motion as the crucial but forgotten breakthrough in ancient ontology, and this understanding not only informed his account of human existence, but led him to reject the Darwinian biology that Kant's endorsement of mechanism underwrote and to explore nonreductive approaches to living and animal being.

Aristotle's influence on Heidegger has been well documented. What has been less noted is the fact that his engagement with Aristotle involved the search for an ontology of life; life not only in the sense of the living, breathing, corporeal human being, but also in the broader sense of animate being as such. Indeed, it is commonly held that Heidegger's thought is inhospitable to life philosophy. Spirited declamations abound. Didier Franck: "The ecstatic determination of man's essence [by Heidegger] implies the total exclusion of his live animality, and never in the history of metaphysics has the Being of man been so profoundly disincarnated."[1] Derrida: "[T]he distinction between the animal and man has nowhere been more radical nor more rigorous than in Heidegger."[2] Agamben: "[Heidegger] is the philosopher of the 20th century who more than any other strove to separate man from the living being."[3] Yet as David Farrell Krell notes, "However much Heidegger inveighs against life philosophy, his own [early] fundamental ontology and [later] poetics of being thrust him back onto [it] again and again."[4] While Heidegger has long been cast as hostile to or neglectful of life philosophy, his work on Aristotle in the 1920s demonstrates a struggle to articulate an ontology of life.

This is no peripheral concern in his work and should be seen in the broader context of the development of his philosophy of nature. Indeed, by the 1930s, the question of being had for Heidegger become all but synonymous with the question of nature. Heidegger's mature philosophy of nature is an answer to scientific naturalism, and a key part of this conception is the ontological status of life. In *Fundamental Concepts of Metaphysics* (hereafter *FCM*), Heidegger analyzes the concept of world in part through a phenomenological investigation of life and the animal. In doing so, he draws on research in zoology and biology that he sees as spearheading a revolution in the understanding of life. He detects in the biology of his own day

> a fundamental tendency to restore autonomy to "life," as the specific manner of being pertaining to animal and plant, and to secure their autonomy for it. This suggests that within the totality of what we call natural science, contemporary biology is attempting to defend itself against the tyranny of physics and chemistry. . . . The task confronting biology as a science is to develop an entirely new projection of the objects of its inquiry. . . . [and] to liberate ourselves from the mechanistic conception of life.[5]

He is likely alluding to Uexküll, who here contrasts biology with physiology:

> For the physiologist, every living thing is an object that is located in his human world. He investigates the organs of living things and the way they work together just as a technician would an unfamiliar machine. The biologist, on the other hand, takes into account that each and every living thing is a subject that lives in its own world, of which it is the center. It cannot, therefore,

be compared to a machine, only to the machine operator who guides the machine.⁶

The physiologist—or behaviorist or scientific naturalist—does not appreciate what we might call both ontological and methodological pluralism. The intentional comportment and interests of the researcher condition what will and will not show up, and how it will show up, on his radar. Uexküll uses the example of a researcher of airwaves and a musicologist; the former encounters waves, while the latter encounters tones. But both are enacting the same object, and "both are equally real." So multiple methodologies can disclose the same object in different ways; the meaning or sense of a thing is dependent, at least to some extent, on the context in which it is being disclosed. Yet Uexküll appreciates the difficulty of integrating these different perspectives into a unified whole:

> The role Nature plays as an object in the various environments of natural scientists is highly contradictory. If one wanted to sum up its objective characteristics, only chaos would result. And yet, all these different environments are fostered and borne along by the One that is inaccessible to all environments forever. Forever unknowable behind all of the worlds it produces, the subject—Nature—conceals itself.⁷

This passage draws out the way in which Uexküll's approach is neo-Kantian. He held that we cannot know Nature in itself; the "plan" of nature—its ultimate order and structure—eludes our grasp but subtends all the phenomena that do show up; how similar this is to Heidegger's notion of being as simultaneously revealing and concealing. We will see, in future chapters, how this notion of nature echoes in Heidegger's later work, though the concepts of *physis* and earth. But Uexküll is also neo-Kantian in how he extends Kant's thinking about space and time to organisms: "Without a living subject, there can be neither space nor time. With this observation, biology has once and for all connected with Kant's philosophy, which biology will now utilize through the natural sciences by emphasizing the decisive role of the subject."⁸ The error of the reductionist is thus to naively employ concepts that derive from and depend on the lifeworld, and "flatten" the different orders of being into one: the physical or physiological. So Uexküll's *Umwelt* theory embraces on ontological pluralism: there really are different orders or levels of reality in nature. While our access to them is limited (i.e., we cannot know exactly what it is like to be a bat), we must acknowledge that they are there: that there is *something like* what it is to be a bat that cannot be fully explained or understood in physiological terms; and, moreover, that we can gain a better grasp of it through a combination of observing the organism's behavior and examining its physiological structure in order to correlate the latter with certain experiential potentials.

But back to *FCM*: like Uexküll, Heidegger appreciates the importance and the difficulty of working out a "metaphysical interpretation of life" and alludes to the "inner unity of science and metaphysics."[9] These formulations are tantalizing, but Heidegger never follows through on these promissory notes. One reason he does not, I think, is because he is captured by (and never escapes) Kant's conviction that biology cannot be a science because it traffics in regularities, not certainties. As Glazebrook has shown, for Heidegger, science is essentially the mathematical projection of nature. Earlier on, and still in *FCM*, he appears to hold out hope that philosophy can both ground the sciences (through regional ontologies) and be the master science, with ontology as the science of being. In this way, he rejected the metaphysical neutrality of Husserlian phenomenology. As Glazebrook summarizes,

> By using phenomenology as a scientific method for doing ontology, Heidegger rejects the bracketing of metaphysical issues for which Husserl's phenomenology called. He accepts Husserl's conception of regional ontology, in which the sciences define some realm of beings as their object by projecting a basic concept. But Heidegger further argues that metaphysics, in contrast to the sciences, takes being as its object.[10]

In other words, Heidegger initially thought that phenomenology would lead to a fundamental transformation in our understanding of science. The problem, however, is that he realizes that since the basic concept projected by biology, namely life, cannot properly be explained in mechanistic terms, biology is not really a science. Put differently, the projection of biology (life) is incompatible with that of science (the mathematical projection of nature). What is more, later on he becomes convinced that science is, as it were, teleologically technological: nature is mathematically projected in order to be technically manipulated. These connections will become clearer in the following chapters. But before moving on, I do want to point out that a "fundamental transformation of science" is precisely the possibility opened up by Uexküll's expansion of phenomenology into biology and pursued by his intellectual heirs in the field of biosemiotics, which John Deely deems "perhaps the most international and important intellectual movement since the taking root of science in the modern sense in the seventeenth century."[11] This was Uexküll's conviction: that just as Husserl and phenomenologists saw phenomenology as more truly empirical than an empiricist epistemology and scientific naturalism, a biology that took into account the perceptual and purposive dimensions of living things would be more evidence based.

For now, the point is to see that in *FCM*, Heidegger is stuck. He wants to avoid both positivism and anthropomorphism; indeed, in a telling admission in his later Heraclitus course, he says that "human analysis practically runs out of alternatives when it rejects mechanistic views of animality . . . as firmly as it avoids anthropomorphic interpretations."[12] In trying to thread this needle, Heidegger was arguably pursuing what today goes by the name of a nonreductive naturalism, which would include the goals for the philosophy of biology cited above—to "restore autonomy to life" and "develop an entirely new projection of [biology's] objects of inquiry." These 1929–30 investigations are tantalizing, but they are widely thought to be the closest Heidegger gets to carrying out such a project.[13]

However, an alternative naturalism is precisely what Heidegger explores in his studies of Aristotle in the early 1920s. In these early lecture courses, Heidegger appropriates Aristotle's key concepts of life, *psyche*, *kinesis*, and *physis* in order to sketch an ontology of life. At this point in his thought, animate life—and not in the merely human sense—is the pivot of the being question. All told, Heidegger's early ontology of life paints a more continuous view of the human-nature relationship than his middle and later writings, one in which something like soul spans the spectrum of living things. Yet after the late 1920s, he never returns to this theme in great detail. While he is correct to contest scientific or reductive naturalism, his investigations into theoretical biology in the late 1920s could have led him to a nonreductive form of naturalism, rather than the poetic view of nature that dominates his later writings.

Pre-*Being and Time*: Aristotle on Life, Soul, and Nature

Heidegger's view of life is distinct from his view of animality. When he speaks about life, he is usually referring to "factical life" or human existence, a notion he believes must be clearly separated from any notion of animality. When Heidegger does discuss animality, he does so more in the spirit of the Aristotelian tradition (of seeing living beings as "animated") than the modern tradition (which views them as mere physiological beings). Indeed, As Krell notes, "animation . . . is Heidegger's principal preoccupation both before and after *Being and Time*, from the period of his hermeneutics of facticity to that of his theoretical biology."[14] There are roughly three phases to Heidegger's thinking about animals: 1) Before *Being and Time*, Heidegger weaves together elements from the biological works of Aristotle and Jakob von Uexküll into an ontology of life that ascribes disclosedness, being-in-the-world, being-with (*Mitsein*), and the capacity for signification to animals. 2) In *Being and Time* and *FCM*, he claims that they have an *Umwelt* but, as *FCM* puts it, are "world-poor," bound by their drives. 3) In *Introduction to Metaphysics* and after, he insists that animals have *neither* world *nor* environment and generally seems to lose interest in the concept of the animal. All told, despite Heidegger's promising pre-*Being and Time* sketches for an ontology of life and his later occasional, cryptic, and intriguing comments to the contrary, his position roughly from *Being and Time* onward is that there is an essential, ontological separation between human and animal being, one that should not primarily be regarded as a difference in species. It may be "nonreductive," but it can no longer be called any kind of "naturalism."

In seizing on the peculiar movement of prereflective factical life, Heidegger believed he was uncovering a stratum of being long neglected by the tradition that had first been worked over by Aristotle, and his creative appropriation of this stratum would lead to his famed conception of being-in-the-world, which was developed in close concert with considerations on life and animality. While developing his conception of being-in-the-world over a series of early lecture courses that mainly deal with Aristotle,[15] Heidegger carries out sustained analyses of life and animal being. Here, he attributes much more to life and animal being than is generally done both in modern philosophy and biology and in his own later work. In some of the first of these lecture courses, in 1921–22 and in a 1922 essay, Heidegger provides sketches for an extensive book on Aristotle that would treat the following themes: "The problem of beings and the sense of being (*on—ousia—kinesis—physis*) [beings—Being—movement—nature]."[16] In the 1922 essay, he states that the goal of the book is to offer a new ontology by tracing the meaning of being down through increasingly more fundamental levels: from Being, to human Dasein, to life, to *physis*. The first part of the book was to focus on the *Nicomachean Ethics*, the *Metaphysics*, and the *Physics* in order to discern the nature of human being. But in the second part, Heidegger announces that the foundation for the latter is an ontology of life:

> Aristotle's ethics is then to be placed into this ontological horizon [of the first part], such that this ethics is seen as the explication of beings in the sense of human beings, i.e., human life and its movement. This is done in such a way that we first provide an interpretation of *De Anima* . . . and indeed this itself is carried out on the broader basis of an explication of the domain of the being of life as a particular kind of movement (i.e., on the basis of an interpretation of *De Motu Animalium* [On the Motion of Animals]).[17]

The point is that Heidegger approaches ontology in terms of *movement* rather than *things*, that the main movement he is concerned with is that of life, and that early on, life is conceived not merely as the temporal movement of human Dasein, but in a wider sense to include both humans and other living things. Here, it appears that Heidegger is claiming that an ontology of life is more fundamental than what he will later call "fundamental ontology," that is, the ontology of human existence. Though he will later retract many of these ideas, it is telling that his early experiments in fundamental ontology explore an ontology of life in such detail.

In the Aristotle lectures from 1921–22, Heidegger devotes the third part to an exploration of "factical life." While the sense of life targeted in this lecture is predominantly human life, Heidegger's fledgling forays here, guided by Aristotle's definition of life as self-moving, set the stage for his theoretical biology of 1929. He begins his 1921–22 course by laying out three theses about life. First, it has a temporal cohesion that is bound and finite. Second, this cohesion consists of a set of possibilities. Third, these possibilities can be developed "from within," and they can befall life "from without." Krell summarizes the three theses: "The whole of life, as the temporal process of a bounded stretch of possibilities that we shape and that shape and befall us, is called actuality, *Wirklichkeit*."[18] Life is thus being approached in an ontological register, as a kind of reality; indeed, Heidegger specifically identifies Aristotle's approach to life as putting ancient ontology on the proper path.

Two important aspects of this reality intrigue Heidegger: its peculiar movement and its inherent relationality. First, he characterizes movement as an intentional directedness toward the world from potentiality to actuality. Hans Jonas nicely captures this aspect of Aristotle's view. The temporal structure of life, Jonas argues, suggests a teleological interpretation. Life, he says,

> is essentially also what it is going to be and just becoming: in its case, the extensive order of past and future is intensively

reversed. This is the root of the teleological or finalistic nature of life: finalism is in the first place a dynamic character of a certain mode of existence, coincident with the freedom and identity of form in relation to matter, and only in the second place a fact of structure or physical organization.[19]

To be sure, Heidegger eventually reverses Aristotle's view of the actuality/potentiality relationship and rejects the doctrine of substantial forms mainly because he is convinced they depend on a prior interpretation of being as presence. In his 1926 lecture course on ancient philosophy, he notes that

[for Aristotle,] *dynamis* [potentiality] and *energeia* [actuality] are two basic modes of presence-at-hand, of *ousia* [substance]. Thus they refer back to genuine Being, the Being of the categories. *Energeia* is the highest mode of Being. *Energeia* is prior to *dynamis*, "actuality" before "possibility": to be understood on the basis of the fact that Being means presence.[20]

Heidegger sees Aristotle's discovery of these categories as a breakthrough in ancient ontology concerning the nature of motion. His interpretation of *dynamis* as a term for the unique possibilities proper to a being that determine its range of projects is the prototype for his notion of readiness-to-hand. As he puts it in the 1922 essay, Aristotle helps us see how "objects are given in terms of their full significance in the environing world" and as "being-found-along-with."[21] He insists that "the fact that Aristotle was able to bring this being-found-along-with into relief as a separate sense of being is at the same time the strongest expression of the fact that he did take up the environing world as it is fully experienced."[22]

While Aristotle's categorial breakdown of the being of life is the clue to a better interpretation, Heidegger thinks his view is skewed because of its prioritization of the present. The *anima* of life is a kind of restlessness; it is driven out of itself in the sense that its self, its being, is undetermined by any principle of form. It is not as though there is first the self-contained living being that at some point goes beyond itself into the *Umwelt* or world; rather, it is always already entangled with its environment. Central to Heidegger's approach here is the notion that by shifting our focus from the individual entity, as an extant, isolated substance, to its environmental or worldly situation, as a dynamic, temporally unfolding process, we gain a deeper insight into its being.

The second aspect of life Heidegger seizes upon is that it is inherently relational: it is always already referred to a world or, as he puts it a few years later, is being-in-the-world. Here we see Heidegger distancing himself from what, presumably, he regards as Aristotle's vitalism, or at least what has come to be interpreted as Aristotle's vitalism: the view that living things possess a kind of form or entelechy that causes them to be what they are and develop as they do. Life is constituted as much by its environment as by its own formal possibilities. As Krell notes, "Heidegger's discovery is that—Aristotle to the contrary notwithstanding—factical life is not self-moving. . . . Life needs the security of possibilities that are already 'lived-in,' and that it tends to fixate."[23] The movement of life is just as much a self-being-moved as it is a self-moving.

Heidegger's most direct engagement with Aristotle's writings on life and soul, especially the *De Anima*, is found in the 1926 lecture course *Basic Concepts in Ancient Philosophy*, where he insists that Aristotle's key ontological innovation in the philosophy of nature is his understanding of *zōē* (life). Indeed, Heidegger here presents Aristotle's biology as the basis for understanding Dasein's way of being as a specific kind of living being. *Zōē*, he writes, has an "exemplary significance," it is "the first-ever phenomenological grasp of life," and it "led to the interpretation of motion and made possible the radicalization of ontology."[24] Heidegger outlines Aristotle's *De Anima* by noting the different levels and essential features of life. Soul, he insists, should not be seen, as it commonly is in modern thought, as merely "psychological," but as pertaining to life as such. He interprets perception (*aisthesis*) in terms of world: "*[A]isthesis* . . . discloses the world, though indeed not in speech and assertion, not in showing and making disclosure intelligible. Fundamental concept of sensibility: letting a world be given and encountered by disclosing it."[25] The student transcription of the lecture fleshes this out further: "What is alive, and also stands in a determinate communication with something, is such that it has a world, as we would say today."[26] Heidegger points out that this interpretation of soul/life is the foundation for determining the being of Dasein, which Aristotle addresses in the *Ethics* and *Politics*; so while human being is distinct from animal being, the two are situated much more closely along a common continuum than in Heidegger's later writings.

The other major influence on Heidegger's early explorations of life is the biologist Jakob von Uexküll. Indeed, Uexküll's insistence that the organism can only be understood in relation to and oriented toward its environment seems to have influenced his reading of Aristotle. Noting that

Aristotle is assumed to be the father of the "theory of the soul as a substance," Heidegger asks:

> But what if [this assumption] rested on a fundamental misunderstanding of the sense and intention of the Aristotelian theory of the soul? There is so little of the soul as a substance, in the sense of a physical breath, housed for itself somewhere in the body and at death vanishing into the heavens, that it was precisely Aristotle who first placed the problem of the soul on its genuine ground.[27]

So rather than see the soul as something "in" the organism, Heidegger suggests that Aristotle sees the soul as the power or potential for comportment toward a world, as a kind of intentionality.

Much like Aristotle, Uexküll practiced biology primarily through field research, studying animal behavior in natural environments. For Uexküll, the implications of *Umwelt* research for understanding life were radical: his idea was that each animal inhabited a kind of soap bubble, an inner world, something like a first-person perspective. Uexküll means "inner" in the sense of inhabiting a horizon of meaning, not in the sense of locked within its own "mind" that represents a pregiven world. As Brett Buchanan explains, "The *Umwelt* forms a figurative perimeter around the organism, 'inside' of which certain things are significant and meaningful, and 'outside' of which other things are as good as nonexistent insofar as they are 'hidden in infinity.'"[28] Animals enact and bring forth their environment; there is not just one region called the environment that is perceived differently by different organisms.

Uexküll's theory not only claimed a sphere of interiority for animals, but also embraced what Buchanan calls an "intersubjective theory of nature."[29] For Uexküll, animals do not merely perceive and react; they interpret and respond in a novel way to both their physical surroundings and other organisms. Buchanan notes how Uexküll's ideas have been taken up by thinkers in the field of "biosemiotics." Jesper Hoffmeyer, for example, pushes intentionality all the way down, linking it to bodily movement and activity: "[E]ven amoeba, [Hoffmeyer] wants to say, anticipate their surroundings by interpreting cues and signs as meaningful, and thus suggest a kind of intentionality toward their *Umwelt*, no matter how innocent and rudimentary this may be."[30] On this view, organisms must not be seen merely as passive objects in systems of causal interaction, but as interpreting subjects in networks of communication that can generate and register signs in their environment.[31] And something like this view is just what Heidegger embraced in the early 1920s.

Drawing on his early sketches for a neo-Aristotelian ontology of life, Heidegger makes extended comments on life and animality that draw on examples from Uexküll's research on animal *Umwelts*. For example, Heidegger concedes more to animal being in the 1924 lecture course, *Basic Concepts of Aristotelian Philosophy*, than perhaps anywhere else in his oeuvre. Here, animals are said to signal and indicate, to have *Mitsein* (being-with), and to possess world; he explicitly refers to "the being-in-the-world of animals."[32] Discussing Aristotle's treatment of "speaking" in the *Politics* and *Rhetoric*, Heidegger says of animals, "The being-possibility of animals has of itself reached this mode of being [i.e., "speaking-about"], having perception of what constitutes well-being and being-upset, being oriented toward this and indicating this to one another."[33] He specifies different kinds of animal indication a few pages later:

> Enticement and warning have, in themselves, the character of addressing itself to [something]. . . . Enticing means to bring another animal into the same disposition. . . . [These] have in their ground being with one another. Enticing and warning already show that animals are with one another. . . . Since animals indicate the threatening, or alarming, and so on, they signal, in this indicating of the being-there of the world.[34]

While he qualifies this by noting that human speaking is distinct in that, since it involves *nous* (reason) and *logos* (discourse), it can identify what is "good and evil" or "proper and improper" and thus can serve as the basis of household and political community, the comments are striking in their characterization of animal being in terms of possibility rather than actuality, as having a capacity for meaningful communication, and as possessing being-in-the-world. Granted, Heidegger will later see a deeper rift between *semantike* (meaning), which involves *nous* and *logos*, and *phone* (sound), which does not, but here he nevertheless does ascribe a more robust semantic sphere to the animal.[35] Indeed, his language here sounds very similar to the "hierarchy of souls" view in the *De Anima*: there are, he says, "different gradations and levels" of world disclosure between humans and animals.[36] It is exactly this *scala natura* language that Heidegger will eschew after *Being and Time*.

In a 1925 lecture on Dilthey, Heidegger makes some striking remarks about animal and living worlds:

> Life is that kind of reality which is in a world and indeed in such a way that it has a world. Every living creature has its environment not as something extant next to it but as something that is there for it as disclosed, uncovered. For a primitive animal, the world can be very simple. But life and its world are never two things side by side; rather, life has its world. Even in biology this kind of knowledge is slowly beginning to make headway. People are now reflecting on the fundamental structure of the animal. But we miss something essential if we don't see that the animal has a world.[37]

In the same year, Heidegger, presumably drawing on one of Uexküll's examples, uses the case of a snail's being "in" its world as different from being "in" its shell to illustrate his concept of being related to a world.[38] In a 1925–26 lecture course, he even goes so far as to say that plants have a world in some sense, citing precedent and common ground in Aristotle and Karl von Baer: "[E]specially in the 19th century, reference has been made to this structure [of being-in-the-world] . . . to the fact that animals above all, and plants in a certain sense, have a world. To my knowledge the first person to have run across these matters again (Aristotle had already seen them) was the biologist Karl von Baer."[39] A year later, in a lecture on the concept of truth, he will assert that even a jellyfish in some sense discloses the world and, what is more significant, he relates this to the notion of Dasein as a "clearing" and as "uncovering," contrasting the two with a nonliving object such as a chair:

> If any being called Dasein or *something living* is, it is in a world. It is on this basis that the doctrine of *lumen naturale* (natural light) must be understood: understood philosophically, we can say that human Dasein has such a kind of being that it bears a light in itself. . . . The chair is in the world in a different way. It does not have what it is in as space. The floor on which it rests is not accessible, disclosed, to it, whereas our way of being is of such a kind that we are, according to our essence, always already in a world. Even a jellyfish has, when it is, its world. Something like a world, a being that it itself is not, is uncovered, disclosed, for it.[40]

Conclusion

While it is clear that Heidegger had not yet fleshed out the differences between *Umwelt* and world,[41] and though there is probably some semantic slippage between the two, the passages cited above are significant because Heidegger goes beyond referring to animals as particular kinds of being-in-the-world and speaks about life as a "kind of reality." In other words, his view implies and demands something like the ontology of life he had promised in 1922. The upshot of his scattered remarks is that, prior to *Being and Time*, Heidegger appears sympathetic to the position that animals have world and that there is some sort of ontological continuity between human and nonhuman life. By situating Dasein in terms of Aristotle's "natural cosmos" rather than what he took to be the transcendental and anthropocentric categories of medieval and modern philosophy, Heidegger's pre-*Being and Time* writings arguably come closer to a nonreductive naturalism than anything else in the rest of his corpus. Why, then, did he eventually abandon this project? To understand why, we need to examine *Being and Time* in detail: its method, structure, and goals, and especially its treatment of the concepts of life and nature.

3

Life and Nature in *Being and Time*

Though nature appears to be a peripheral concept in *Being and Time*, I contend that it exerts a subtle pressure that both frustrates the work's completion and forces Heidegger to reformulate his project; moreover, it is no accident that the concept of life, set to the side at the start of the work, is never fully engaged, and is deferred until *FCM*. Heidegger approaches nature in the way he does for methodological reasons. The conceptions of nature proper to what Husserl called the natural and theoretical attitudes must be "bracketed" or held in abeyance. Heidegger suspends ontological assumptions about nature (and life) in order to clarify how the different senses of nature are founded on and arise out of the modes of human intentionality. Heidegger's sparse discussion of nature in *Being and Time* is found primarily in his analysis of human inauthenticity, our average, everyday way of going about our business and attending to our concerns. This analysis, however, is but a preparation for the pivotal second division in which he famously claims that Dasein's understanding of itself and its world is determined by its finite temporal structure and that the meaning of being, in all its permutations throughout the Western philosophical tradition, has been determined by an interpretation of time first put forth by Aristotle. Thus the conceptual link between being and time is the crux of the text. Nature appears to be ancillary.

One of Heidegger's interpretive principles is that in studying a text we can and should seek to bring to light the "unsaid," that is, what the author does not explicitly *say* but what covertly conditions and quietly pervades what *is* said throughout the work. Heidegger's marginalization of the question of nature in *Being and Time* can be seen as one such "unsaid" that erupts in his later works in the form of a philosophical biology, in 1929–30, and a new notion of earth and a focus on *physis* (the Greek word for nature), in the 1930s and '40s. Hence Heidegger's treatment of the question of nature can tell us much about the development of his thought and the relationship between his early and later work. In order to understand why he becomes preoccupied with these nature-related themes in his later work, we first have to determine how he approached—or why he avoided—the question of nature (and life) in *Being and Time*.

First, I summarize the aim, structure, and method of *Being and Time* in order to convey Heidegger's general philosophical concerns. Second, I trace the emergence of the theme of nihilism in Heidegger's work in and before *Being and Time*. As I show in the following chapters, this theme only grows in importance as his thought (and his conception of nature) develops. Third, I explain in more detail one of the central features of Heidegger's existential analytic: the notion of world. My claim here is that this feature is essential to Heidegger's discussion of nature. Fourth, I address Heidegger's treatment of the concept of life in the text. Fifth, I discuss the three senses of nature addressed in the text by drawing on Bruce Foltz's landmark monograph of Heidegger's treatment of nature, *Inhabiting the Earth*. Finally, I explore problems stemming from Heidegger's account, focusing especially on Hubert Dreyfus' classical commentary on *Being and Time*.

My aim here is to determine what misfires in his early approach to nature and how this prefigures his later formulations. My conclusion is that Heidegger's thinking about nature splits into two tracks after *Being and Time* that are never reconciled. One track leads to *FCM*, in which he analyzes worldhood in terms of not just Dasein but animals and inanimate natural beings and addresses a host of issues pertinent to philosophical biology and ecology. This track leads in the direction of a phenomenological biology or phenomenology of life, a path Heidegger chooses not to pursue. The second track leads to the mainsprings of his later thought, such as the enigmatic notion of earth, his retrieval of *physis*, and his "history of being." While this track allows Heidegger to work out a noninstrumental, nonnaturalistic, and more poetic notion of nature, it leads him to neglect the important issues broached in the former track and to maintain a sharp ontological divide between humans and animals/life/nature, a divide some environmental thinkers find problematic. My point is that this fissure is already nascent in *Being and Time* and is a consequence of his approach of fundamental ontology.

I. *Being and Time* in Brief

The philosophical ambition of *Being and Time* is grand. Heidegger begins the treatise by pointing out that the question of being has been forgotten, trivialized, and taken to be superfluous. He highlights the peculiarity of being, noting that it is at once the most universal, indefinite, and self-evident concept. But rather than see these as reasons to pass over the question of being, Heidegger is convinced that they should lead us to confront it head on, since we presuppose some understanding of the word in all that we do, say, and think.

Heidegger is convinced that the prejudice that the question of being is a nonstarter is "rooted in ancient ontology itself."[1] Hence, right at the start of the text, he posits a connection between the present understanding of being and that of the ancient world. As such, *Being and Time* offers not just a critique of the Western philosophical tradition, but an account of human existence. The connection between these two is being. The reason this connection has remained hidden, Heidegger contends, is that the meaning of being is determined by time. As he explains in the second introductory chapter,

> time needs to be explicated primordially as the horizon for all understanding of being, and in terms of temporality as the Being of Dasein, which understands Being. This task as a whole requires that the conception of time thus obtained shall be distinguished from the way in which it is ordinarily understood. This ordinary way of understanding it has become explicit in an interpretation precipitated in the traditional concept of time, which has persisted from Aristotle to Bergson and even later.[2]

The tradition has failed to appreciate how it has been determined by previous interpretations of being and how these interpretations have been determined by a sense of time that arises out of everyday experience. There is thus what Heidegger calls the "ontic priority" of the question of being: it must depart from the way being is understood by one particular kind of being, human being. This is why he thinks that the question of being should be broached at the intersection of everyday experience and the philosophical tradition.

Heidegger tells us that *Being and Time* must accomplish two tasks in order to address the question of being: provide an exhaustive analysis of human Dasein and a destruction of the history of ontology. The treatise is thus divided into two parts. The first part Heidegger calls "the Interpretation of Dasein in terms of Temporality, and the Explication of Time as the Transcendental Horizon for the Question of Being." This in turn is split into three divisions, of which only the first two were completed: 1) the analysis of Dasein as being-in-the-world, 2) temporality as the meaning of Dasein's being, and 3) time as the conceptual key to the question of being.[3] The point of departure for this "existential analytic" is what Heidegger calls "average everydayness," the condition in which we find ourselves always already involved in a world and in which we possess a "vague average understanding of being" prior to any theoretical separation between subject and object.[4] The destination is the discovery of how our everyday understanding of being is determined by an ordinary conception of time, which is defined in terms of the present, but is actually founded upon a prior conception of time, which issues from the future and is grounded in human mortality. This more primordial conception of time reveals both Dasein's mortality and its original access to being, that is, the basis of its ability to do ontology. As such, this deeper understanding of time is to be the key to resuscitating the question of being that has been passed over and neglected throughout the tradition because the latter based its understanding of being on the ordinary conception of time.

The existential analytic of part 1 unfolds in two acts. The first division is wholly preparatory in that it offers an ontological interpretation of Dasein in its everydayness. When Heidegger claims at the outset that "the essence of Dasein lies in its ex-istence," he means that the a priori constitution of Dasein is presupposed and operative in its concrete being in the world.[5] As such, Heidegger's phenomenology aims to analyze all of the modes of this concrete being in order to distill the basic structures or categories that make the latter possible. Moreover, the analysis is not intended to be a general account of the human species, but must be seen as "in each case mine."[6] The first-personal dimension of the phenomenon in question, Dasein or transcendental subjectivity, must be respected, and this is one of the reasons why Heidegger insists that his investigation is in no way offered as an anthropology, psychology, or biology. It is also why the field of investigation is different from, say, that of Kant's *Critique of Pure Reason*, which is a search for the conditions of the possibility of experience for subjects in general and is, furthermore, restricted to the mode of intentionality concerned with representation and objectification. The existential analytic is, however, unmistakably patterned on Kant's critique: it is a transcendental inquiry into the universal and necessary conditions for the possibility of doing ontology. The question is not how synthetic a priori judgments are possible (how humans are able to cognize things about the world in advance of experience), but how any understanding of being at all is possible (how humans have a world). Thus Heidegger shifts the site of analysis from the theoretical judgment to the pretheoretical understanding of being, which is always being *in a world*. Heidegger calls this our "pre-conceptual understanding of being," that is, our ability to differentiate between being itself and particular beings, our capacity to encounter beings *as* beings, to be in a world, to be historical, and to have language. The structure to be distilled—the existential constitution of Dasein—is an a priori whole that is presupposed in its entirety no matter the aspect by which it is being approached.[7]

The task of the first division of *Being and Time* is to show that this complex of intentional structures is "being-in-the-world" and that its meaning is "care." Heidegger begins this analysis by first breaking down the various senses of world and showing how they are founded upon this basic structure. The primary way in which beings show up for us is in their "readiness-to-hand" (*Zuhandenheit*) in our environment, as equipment available for completing certain tasks we assign ourselves. Our discovery of and encounter with things in this mode of being is pretheoretical, and the sense of these things is determined by their place and function within a purposive context. Only when this context is ruptured, when something disturbs the smooth operation of our work and frustrates our purposes, do we disengage from our involvements and take up a theoretical stance toward things. Only then do we encounter things as simply there, as bare objects, as "present-at-hand" or merely occurrent. The sense of being as "presence-at-hand" (*Vorhandenheit*) is thus derivative of and founded on the more basic sense of being as "readiness to hand," yet the tradition erred in taking it as primary and thus defining both Dasein and the world ontologically in terms of it. The sense that things within the world have for us is for the most part confined to these two modes. Heidegger distinguishes between these modes of being, which are "categorial," and the modes of Dasein's being, which are "existential."

Heidegger surveys and identifies the whole of the a priori structure he has detailed—"care"—and reveals the condition for the possibility of its apprehension in the mood of anxiety. Anxiety reveals Dasein to itself in a distinctive way because unlike other moods, such as fear, it is not directed to any object within the world. As such, anxiety scrambles Dasein's intentional radar, disrupts its everyday absorption in its world, and reveals its fallenness toward entities. In anxiety, the world, as the frame of reference, horizon of sense, or context of meaning that makes things intelligible, is encountered *as* a frame, *as* a horizon, *as* a context, and this disorients Dasein because it had taken the horizon for granted as a polestar guiding all its activity. When anxiety throws Dasein face to face with its sheer being-in-the-world as a whole, Dasein discovers its ex-istence, that it is always already "ahead of itself" and becomes aware of its own possibilities of being. Moreover, Dasein realizes that this was its ever-present condition—"anxiety is always latent in being-in-the-world."[8] Catching itself in "midfall," it sees that it had always understood itself and its world in terms of its own possibilities, but had failed to own up to this fact. Dasein discovers, in other words, that it is not like either other entities in the world or other Daseins, and that these cannot help it to determine itself. Only because Dasein is ahead of itself, that is, self-transcendent, can it be *for* itself, and only anxiety can manifest this condition, because it reveals Dasein's radical freedom. This is why Heidegger says that anxiety "individuates." Division 1 ends with a conclusion—that the being of Dasein is care—and a question: Is there a "still more primordial phenomenon" that grounds the unity and totality of the care-structure?

The second division begins by announcing that the analysis of the first division is incomplete. The initial analysis fails primarily because it abstracts from Dasein's temporality. Recall that Heidegger's methodology called for beginning with what is most concrete. Initially, it appeared as though Dasein's everyday being in the world was a sure basis from which to begin, since it revealed the phenomenon of world that had been passed over by previous ontologies. However, Heidegger pulls the rug out from under this position by showing that the mundane, quotidian, average everyday comportment toward the world is a kind of false consciousness unmasked by the first part of the existential analytic. The attempt to grasp Dasein as a whole fails because of the mistaken assumption that Dasein *is* in fact whole. The analysis of the first division misunderstood the object it analyzed, precisely by conceiving it *as* an object. The second division, then, is what Hubert Dreyfus, employing Ricoeur's phrase, usefully calls a "hermeneutic of suspicion," which aims to show how Dasein's supposedly concrete, familiar, reliable view of itself and its world is actually an elaborate defense mechanism for covering up its own incompleteness and denying its own death.[9] As such, the first chapter of division 2 takes death as the "end" of being-in-the-world and the key to making sense of its aspiration for completion. As Heidegger puts it, "As long as Dasein is, there is in every case something still outstanding, which Dasein can and will be. But to that which is outstanding, the end itself belongs."[10] The "end" of death must be reconceived not as the cancellation of Dasein's being and actuality, something external to it that is destined to occur at some point in the future, but as pervading Dasein's every mode of being and as the condition and limit of all its possibilities: something internal to it that is, in a sense, already occurring.

Heidegger answers the question he posed at the end of the first division by claiming that temporality is the foundation of care. Since Dasein's authentic potentiality for being a whole consists in an *anticipatory* resoluteness toward death, since this stance clearly has a temporal (specifically, futural) dimension, and since this possibility, encountered in anxiety, lies at the ground of Dasein's average everydayness (the care-structure), it follows that temporality is the condition for the possibility of care. Heidegger connects the care structure with temporality: "Dasein's totality of Being as care means: ahead-of-itself-already-being-in (a world) as Being alongside (entities encountered within-the-world). . . . The 'ahead-of-itself' is grounded in the future. In the 'Being-already-in . . . ,' the character of 'having been' [the past] is made known. 'Being-alongside . . .' becomes possible in making present."[11] Heidegger refers to these three temporal dimensions of past, present, and future as temporal "ecstases" because "temporality is the primordial 'outside of itself' and for itself," and thinks that the ecstatic character of temporality is obscured in everydayness because Dasein comes to understand its being and that of the world in terms of presence-at-hand.[12] Dasein's inclination to interpret the whole of time—past, present, and future—primarily in terms of the present leads it to distort its being, warp its world, and deny its death.

At the conclusion of the book, then, Heidegger tries to determine how this conception of time arises. As thrown into the world, Dasein has to reckon with the time it is given, and since it exists in three temporal dimensions, it is free to emphasize one to the detriment of the others. In going about its business, Dasein uses certain instruments to measure and count time in order to make time public and available. In manipulating its environment, it tends to fix things as present in order to understand them. This activity of "making present" results in an interpretation of time as a sequence of nows moving in a forward direction, in which the past *is* no longer and the future *is* not yet. Over time, this interpretation of time congeals into an ontology that takes all beings as present-at-hand and truncates the fullness of Dasein's temporality. This understanding of time, generated from Dasein's pre-theoretical life, became the traditional conception of time from Aristotle onward, and in Heidegger's view it covertly directed the Western metaphysical tradition. Once all of the presuppositions that Dasein brings to ontological inquiry have been identified, we will be in a better position to determine the meaning of being itself because these presuppositions have purportedly been clouding, coloring, and covering up the understanding of being throughout the entire Western tradition since Aristotle's interpretation of time was locked in as the gold standard and exerted a heretofore unrecognized influence on all subsequent ontologies.

The second part of *Being and Time* was to be what Heidegger calls a "destruction of ontology." What he has in mind is a reinterpretation of the major turning points in the Western tradition of thinking about being in light of the findings of the existential analytic. Since the existential constitution of Dasein is taken to be true for any ontologist, the conceptual edifice of any great thinker can presumably be traced back to and reunited with the phenomenal basis from which it arose. Hence Heidegger planned to move backward through the tradition from Kant to Descartes to the medieval thinkers to Aristotle in order to follow the transformations of basic ontological concepts back to their source. The two main errors in the tradition, he thinks, are the interpretation of time as presence and the interpretation of being as standing presence. These oversights are the consequence of passing over Dasein's being-in-the-world.

Taken together, these two tasks—the interpretation of Dasein and the destruction of ontology—comprise the project of fundamental ontology. The first task pertains more to fundamentals, since it involves identifying the conditions for the possibility of ontology. The second task pertains more to ontology proper. Once the basic structure founding any effort in ontology has been laid bare, the history of ontology can be reinterpreted so that its various conceptual constellations can be refitted to their phenomenal bases. It is important to point out that when Heidegger proposes to "destroy" the history of ontology, he does not have in mind a wholly negative operation of razing the tradition and exposing its basic concepts as bogus and bankrupt. Instead, he sees the tradition as a series of pregnant possibilities waiting to be engaged and brought to fruition. As such, Heidegger conceives the project of fundamental ontology as a completion of metaphysics, not its abolition. *Being and Time* is pitched not as a pox on all ontology, but as the fulfillment of its immanent intention. As Jacques Taminiaux points out, "fundamental ontology attempted to achieve metaphysics by bringing the meaning of Being to conceptual clarity. In the framework of this attempt, the history of metaphysics was not considered an increasing obliteration of Being, but instead as the maturation of the science of Being."[13] If the existential analytic is the "prolegomena," the reinterpretation of the history of ontology is the "future metaphysics." And the link between these two tasks is a critique of the traditional concept of time.

II. Nihilism in *Being and Time*

The term "nihilism" does not appear in *Being and Time*, but its spirit pervades the text in at least two ways. First, in approaching Dasein via the category of possibility, rather than actuality, Heidegger argues that nonbeing permeates Dasein's essence. Rather than see Dasein's movement as the reduction of potency to act in the sense of fixed essence or purpose, Heidegger reconstrues the human "towards-which" as death—an open possibility. Second, his analysis of human existence as fallen, dispersed among *das Man,* and captivated by the world betrays an antimodernism that can be traced back to his early student period. These two forms of nihilism prefigure his later distinction between ontological and historical nihilism, nihilism taken as an essential part of human nature and as a contingent cultural phenomenon. Below I will expound on the second sense mainly as it pertains to Heidegger's presentation of inauthenticity and authenticity, since it sets the stage for his later account. Heidegger was well acquainted with critiques of modernity at an early stage of his career.[14] In a book review from 1910, Heidegger likens an existentialist hero that spurns the comforts and "happiness" of modern life to a "modern day Augustine" who "rests in the shadow of the cross, this strong-willed, joyously hopeful poet-philosopher." Such a "free thinker" "uncovers again and again our great indestructible connections to the past" by resisting the superficialities of the present.[15] Unlike Nietzsche, Heidegger (at this point) saw (conservative Catholic) Christianity as an alternative to modern nihilism, not its root. He began to work the idea of nihilism as fallenness into his sketches of factical life in the early 1920s. In a 1921–22 course on Aristotle, he calls the trajectory of existence "ruinance," a term from medieval Christianity originally tied to sin that connotes corruption. While this prefigures Heidegger's notion of being-toward-death, that concept is strictly demythologized, stripped of any Christian content. It is significant that Heidegger's early thoughts on nihilism are conceived in concert with conservative Christianity's critique of modernity, not in opposition to it; though he comes to reject the Christian content, he retains many of the categories. The framing of factical life as "ruinance," drifting aimlessly toward nothing, will become a mainspring in the concept of inauthenticity.

As I discussed above, inauthenticity for Heidegger means our average, everyday, prereflective way of being and our uncritical acceptance of the beliefs, values, and practices of our culture. Our usual state is one of "fallenness"; thrown into the world, we cling to what is around us and inherit an identity from the people and things around us. These are the materials from which we fashion ourselves, but in inauthenticity, we are unaware of the fashioning, of the constructive moment of our existence; we take it all for granted as "the way things are." In the mode of inauthenticity, Dasein is characterized by *das Man* or "the they," a kind of hive mind in which the individual has not yet differentiated herself from others. *Das Man* shields

Dasein from taking responsibility for its actions and, indeed, from agency itself—it need not interpret its situation uniquely, since the prefabricated script for all situations is readily supplied.

Heidegger considers inauthenticity an essential structure of human existence. This is the most important aspect of the first sense of nihilism, which involves the nonbeing at the heart of Dasein: Dasein cannot avoid falling. Though Heidegger does not see inauthenticity as a contingent historical phenomenon, his account is typical of a reaction against secular humanism, modern liberal values, and bourgeois culture that became widespread during the nineteenth century. His account closely echoes Nietzsche's and Kierkegaard's polemics against the mediocrity of the "herd mentality" and the "leveling" of modern bourgeois culture. The capital complaint of this cast of mind is that most people, most of the time, are not conscious of the meaning, sources, and full consequences of their way of life and make no effort to become so. For Heidegger, the failure to determine who we are is an essential part of what we are. This is very important for his account of nihilism because, as I discuss below, he does not think nihilism (in the above sense of falling into and becoming entangled with things) can be overcome. Indeed, the very attempt to gain complete control over one's destiny, to fully appropriate one's inheritance, to overcome one's finitude and become "whole" and self-sufficient, is the main cause of what Heidegger will call historical nihilism, which reaches its apex in modernity.

Dasein is pulled out of inauthenticity through the experience of anxiety. Anxiety presents a danger and an opportunity: Dasein can either fall back into its old ways or resolve to take up a free relationship to its facticity. That is, it can for the first time consciously and freely bind itself to, own, and take responsibility for its past. This resolution, however, is difficult to make and repeat, and Dasein is so crafty in fleeing this responsibility because it involves a confrontation with death and the recognition that any content we resolve upon has no firm basis. Anxiety shows Dasein that all of the commitments it had taken for granted are not the way things *are*, but merely one way that things *could be*, one way they have been *interpreted to be*. As Heidegger says, we always determine ourselves before nothing: "the 'nothing' with which anxiety brings us face to face, unveils the nullity by which Dasein, in its very basis, is defined; and this basis itself is as thrownness into death."[16] Heidegger's treatment here embraces the idea that the goals, decisions, and projects of the self are not established or justified on the basis of any objective a priori standard, but on the basis of a radical freedom. So as early as *Being and Time* (and before), we can see that nihilism or nothingness plays an important part in Heidegger's thought. I will return to this theme in more detail in the next chapter and link it to his views on life and nature.

III. Heidegger's Understanding of World

At first glance, *Being and Time* seems inimical to environmental philosophy. Heidegger's decision to begin the inquiry into being from the ground zero of *human* existence would seem to speak against the basic intuition fueling much environmental philosophy: that the greatest mistake in Western philosophy is its dominant, if not exclusive, focus on human beings and their interests to the detriment of both other creatures and the earth itself.

The best way to broach the treatment of the concept of nature in *Being and Time* is not to analyze what Heidegger has to say about nature itself, but what he has to say about *world*. For one, the bulk of Heidegger's discussion of nature takes place in the third chapter of Division One, which addresses the worldhood of the world. Moreover, after Heidegger introduces the phenomenon of "being-in-the-world," the first component he addresses is the "worldhood" or being of the world, and the first thing he points out is that the world tends to be taken as "nature":

> [T]o give a phenomenological description of the "world" will mean to exhibit the Being of those entities which are present-at-hand within the world, and to fix it in concepts which are categorial. Now the entities within the world are Things—Things of Nature, and Things "invested with value." Their Thinghood becomes a problem; and to the extent that the Thinghood of Things "invested with value" is based upon the Thinghood of Nature, our primary theme is the Being of Things of Nature—Nature as such.[17]

However, Heidegger insists that a successful determination of nature as such "will never reach the phenomenon that is 'world'" because "Nature itself is an entity which is encountered within the world and which can be discovered in various ways and at various stages."[18]

For Heidegger, world does not mean the purely human world—"culture" as opposed to "nature," a kind of sedimentation of spiritual achievements superimposed on a base of purely natural objects and environments. Similarly, the analyses of temporality and history in the second half of division 2 do not deal with "human history" and "subjective time" in contrast to "natural history" and "objective time."[19] Nor does world mean the sum total of objects over against a subject or community of subjects. As Heidegger bluntly states that "subject and object do not coincide with Dasein and the world."[20] World does not refer to an ontological dimension or place or space different from or separate from human beings. The term Dasein, for Heidegger, is not synonymous with "human beings" or "man" or "an individual human being." As he says, "If we inquire ontologically about the world, we by no means abandon the analytic of Dasein as a field of thematic study. Ontologically, 'world' is not a way of characterizing those entities which Dasein essentially is *not*; it is rather a characteristic of Dasein itself."[21] Worldhood is a part of Dasein's existential constitution and is an essential component of the phenomenon of "being-in-the-world."

Despite Heidegger's tendency to treat nature as a phenomenon founded on being-in-the-world, he at times cryptically alludes to another sense of nature whose being cannot be categorized as readiness-to-hand or presence-at-hand. This sense is nature as an aesthetic phenomenon, "as it is conceived . . . in romanticism" or poetry.[22] Recall Heidegger's stipulation that the "two basic possibilities for characters of Being" are existentials and categories. The latter character of Being applies to things that are present-at-hand ("whats") or ready-to-hand ("what-fors"). The former character applies to entities with the character of Dasein as being-in-the-world—"whos"—on which the categories are ultimately founded. If there is a sense of nature whose being is neither presence-at-hand nor readiness-to-hand (in which case it exceeds categories) nor that of Dasein (in which case it is not an existential), then perhaps Heidegger is inconsistent in claiming that this "romantic" or aesthetic phenomenon of nature "can be grasped ontologically only in terms of the concept of world—that is to say, in terms of the analytic of Dasein."[23] For methodological reasons, Heidegger holds that the being of an entity must be approached in terms of how it shows up for humans, that is, within the world. As Michel Haar observes, "although every being of nature that man encounters—including his own supposed naturality—is necessarily intra-worldly, 'intraworldliness does not belong to the being of nature [itself].' "[24] The latter seems to point to another, transhuman, extraworldly order of being, yet for Heidegger, we can only talk about and conceive of it "in terms of" the world. We can say, then, that his claim is consistent with his methodological constraints. We can say *that* extraworldly nature is, but not *what* it is. Though this recalcitrant, residual aspect of nature escapes the ambit of world, eludes the grasp of Heidegger's fundamental ontology, and, I argue in chapter 5, ultimately leads to his enigmatic notion of earth and focus on the Greek concept of *physis* in the 1930s, suffice it for now to say that, as far as *Being and Time* goes, any analysis of nature must begin with an analysis of world.

In chapter 3 of division 1, Heidegger distinguishes four senses of world. These four senses can be classified in two ways: as ontical or ontological and as existential or categorial. They can refer either to entities (ontical) or to the being of those entities (ontological), and they can refer either to entities with the character of Dasein (existential) or to entities with the character of presence-at-hand or readiness-to-hand (categorial). Heidegger is convinced that the fourth sense, which is ontological and existential, has been passed over by the philosophical tradition and the natural sciences and that this is one of the main reasons for Dasein's forgetfulness of being and misunderstanding of itself. His aim, then, is to show how the other three senses of world are founded on and derived from this primordial sense, the "worldhood" of the world.

The first sense of world is ontical and categorial and "signifies the totality of those entities which can be present-at-hand within the world."[25] Let us call this the "objective world." Propositions about the objective world operate on the level of fact, for example, "the world has seven continents" or "there are no purple cows in the world." Thus there is no concern here for the being of the entities in question—continents and cows—only for facts relating to them. The second sense of world is ontological and categorial and "signifies the Being of [entities that are present-at-hand]."[26] This refers to a determinate realm of certain kinds of objects, objects with a certain kind of being. Taking our previous examples, we could distinguish the world of geology (proper to the continents) and the world of biology (proper to the cows) as distinct from the broader, ontical sense of world as the totality of objects simply present-at-hand. Heidegger is convinced that this second sense of world is the one that has been dominant in the philosophical tradition, has covered up the deeper sense of world, and has masqueraded as "Nature." These first two senses of world are both blind to the structure of being-in-the-world: "Neither the ontical depiction of entities within-the-world nor the ontological interpretation of their being is such as to reach the phenomenon of world."[27] Nor can they properly address the being of Dasein. That is because they only refer to entities *within* the world, which is to say, entities that cannot *have* a world because they lack understanding.

The third and fourth senses of world are accessed through deeper reflection. Like the first sense, the third sense is also ontical and refers to "that 'wherein' a factical Dasein as such can be said to 'live,' . . . [It] may stand for the 'public' we-world, or one's 'own' closest (domestic) environment."[28] The environment is "that world of everyday Dasein which is closest to it."²⁹ It is the arena in which Dasein carries out its concernful dealings. Thus the primary mode of intentionality in the environment is circumspection, and things show up as ready-to-hand. Heidegger is careful to distinguish his sense of the term "environment" from that found in biology. As a positive science, Heidegger claims, biology uncritically operates within the limits of the first and second senses of world, which interpret beings as present-at-hand. In showing that the presence-at-hand of entities is founded on their readiness-to-hand, he argues that the biological notion of environment is unfounded and that it ultimately derives from a more basic sense of the environment as the realm of everyday Dasein's concernful dealings. Put another way, the third sense of world is existential, not categorial: an animal is "in" a field in a different way than Dasein is "in" an—or, more correctly, "*has its*"—environment. What Heidegger means by the term "environment" is markedly different from what many ecologists, environmental philosophers, and ordinary people mean by the term. His point is that when most people *think* about it, they take the environment in the first or second sense of world discussed above: as the totality of birds, bees, and apple trees, as the biospheric whole of which we are parts.

In contrast to the biological sense of environment, this third sense of world as environment refers to our everyday dealings with others and with things, our habitual use of tools to accomplish certain tasks, our "going around" doing something, being involved in definite projects, and so on. The environment is the "around-world" (*Umwelt*), the constellation of people, places, and things in which we go around working out our time and negotiating our tasks, not the container of geometrical or even geographical space through which we move on our way to interacting with various objects. When absorbed in its environment, Dasein does not encounter things as just "there" in their bare physical reality, but discovers them as equipment "referred" to a certain task within a totality of significance or meaning (*Bedeutung*). Where we normally find ourselves—or rather, before we have found ourselves—is as engaged in some task.

But we are not exclusively engaged with the tool or object before us. Our attention is dispersed throughout the meaningful web of instrumental references. It is inclined toward our present project: "To the Being of any equipment there always belongs a totality of equipment, in which it can be this equipment that it is. Equipment is essentially 'something-in-order-to.' . . . In the 'in-order-to' as a structure there lies an assignment or reference of something-to-something."³⁰ This does not mean that Dasein thematizes this totality of significance or even the individual parts of which it is composed. Dasein's tendency is to absorb itself in its practical goal in such a way that neither the tools serving this goal nor the entire referential totality is thematized. As Heidegger puts it, "The peculiarity of what is proximally ready-to-hand is that, in its readiness-to-hand, it must, as it were, withdraw in order to be ready-to-hand quite authentically. That with which our everyday dealings proximally dwell is not the tools themselves [but the work]. . . . The work bears with it that referential totality within which the equipment is encountered."[31] The being of entities in this sense of world is readiness-to-hand, while the being of entities in the first, ontical sense of world is presence-at-hand.

Heidegger's strategy here is to "bracket" the first and second senses of world and perform a phenomenology of the environment in order to determine its ontological structure: the worldhood of the world. This is the fourth and pivotal sense of world, and it is the condition for discovering the being of both the ready-to-hand and the present-at-hand; indeed, it is the condition for our having any sense of anything *as* anything at all. The fourth and final sense of world normally only becomes thematic to everyday Dasein when there is a breakdown in the referential totality, when the engine of our involvement with and absorption in entities suddenly stalls. As we navigate the environment, a foundational horizon of referential meaning is always already presupposed, though it is rarely, if ever, attended to, since it is not itself an object, but rather an intelligible clearing in and through which objects can show up. In Heidegger's terminology, the worldhood of the world is always already *disclosed,* and it is the basis on which entities can be *discovered* (or covered up).

It is imperative to note that Heidegger seizes upon the notion of the worldhood of the world as a third mode of being that is radically distinct from readiness-to-hand and presence-at-hand. Moreover, Heidegger's description of this third kind of being is almost entirely bereft of any reference to nature except for a few cursory references to Dasein's having a "bodily nature."[32] The questions of the "body," "life," and "natural history" are all bracketed in the existential analytic because these concepts are laden with unfounded meanings drawn from the interpretations of nature that stem from the philosophical tradition and the natural sciences. In *Being and Time*, then, natural being is treated primarily as a derivative concept ontologically distinct from human being, and this is so because of Heidegger's understanding of world as the basis of any and all senses of nature. Let us now take a closer look, first, at his treatment of the concept of life, and second, at the different senses of nature operative in the text.

IV. The Concept of Life

There are four points regarding life in *Being and Time* to note. First, the relationship of philosophy to biology (and the positive sciences generally) and the role phenomenology plays in grounding the latter; second, Heidegger's continued reliance on the concept of *Umwelt*, taken from Uexküll, which continues to be a key feature of his description of Dasein's "average everydayness," its being-in-the-world, and of beings as "ready-to-hand"; third, the inadequacy of *Lebensphilosophie* and Husserlian phenomenology; and fourth, the inability of fundamental ontology to access or address the being of life, which escapes the three ontological categories of the text (readiness-to-hand and presence-at-hand, for entities, and existentiality, for Dasein). All of these indicate a turn away from the ontology of life sketched in the previous chapter and toward a more transcendental approach that abandons any biological foundation.

My analysis here is meant to show that life is an anomaly that haunts the text and leads to the theoretical biology of 1929–30. Indeed, at the outset of the latter text, Heidegger notes that he is primarily interested in further clarifying the concept of world; in *Being and Time*, this was done via an investigation of everyday human existence, but in *FCM*, it is done through a comparison of human, animal, and inanimate being. In other words, one of the reasons *Being and Time* is incomplete is because it does not address the ontological problem of life as promised and partly completed in the pre-*Being and Time* writings. As I explain in chapter 5, I see this as the key to Heidegger's difficulties over the concept of nature: he cannot determine the relationship between human and animal being in particular and humanity's place in nature in general except in a "privative" manner; that is, by imaginatively stripping away human reality, which leaves life, animality, and nature as noumenal "things-in-themselves" without any sense or definition. In short: the Kantian strain overcomes the Aristotelian strain.

First, Heidegger originally envisioned *Being and Time* as not merely a critique of metaphysics, but its completion. The introductory chapters of the text make clear the systematic intent: Heidegger insists that phenomenology will construct the mansion, the wider ontological framework, to house all the sciences; it will delineate the different regions of being, determine their proper objects, their methodological limitations and, in general, "put them in their place." Any discipline that oversteps its regional bounds ceases to be science and becomes an -ism, such as psychologism, physicalism, or biologism.[33] As such, this project is more ambitious than the earlier planned book on Aristotle, since it seeks to articulate the division of labor of all the sciences, including biology, and holds that none of them can contribute anything significant to ontology. In his pre-*Being and Time* phase, Heidegger held open the possibility that some sort of philosophical biology was a necessary condition for ontology, but here, that idea seems to be abandoned. Brett Buchanan nails it: "[In *Being and Time*,] Heidegger decisively cuts off further investigation into the anthropological, psychological, or biological sides of human existence."[34]

Heidegger sees the sciences approaching a crisis in their foundations, when the "field propositions" or "basic concepts" are called into question, what we today would call a paradigm shift. For biology, the *Ur*-concept is life: "In biology there is an awakening tendency to inquire beyond the definitions which mechanism and vitalism have given for 'life' and 'organism,' and to define anew the kind of Being which belongs to the living as such."[35] Heidegger appears here to envision a mutually enriching relationship between biology and philosophy: that the former will become more ontologically sophisticated, and that the latter will incorporate key insights from biology, insights to which it may be constitutionally blind. If fundamental ontology misses something important about Dasein, namely its being-alive, and if biology can contribute to the understanding of the being of life, then it follows that biology must be incorporated into fundamental ontology. The latter, then, would no longer be focused exclusively on human existence, but would deal with the living as such.

Whatever the implications of this idea, Heidegger sharply distinguishes his analytic from any kind of biology, unlike in his previous work:

> biology as a "science of life" is founded upon the ontology of Dasein, even if not entirely. Life, in its own right, is a kind of Being; but essentially it is accessible only in Dasein. The ontology of life is accomplished by way of a privative interpretation; it determines what must be the case if there can be anything like mere-aliveness. Life is not a mere Being-present-at-hand, nor is it Dasein. In turn, Dasein is never to be defined ontologically as life (in an ontologically indefinite manner) plus something else.[36]

Heidegger explains the meaning of this passage in a 1925–26 lecture course on logic. In his earlier work on Aristotle, he seemed to approve of Aristotle's "bottom-up" approach to ontology, which regarded the human being as life "plus" something else, that is, *nous* and *logos*. However, after 1925–26 and in *Being and Time*, Heidegger changes his position because of a deepening concern about proper methodology that may be due to the influence of Kant and Husserl; concerned to not let reason overstep its bounds and to deploy concepts responsibly, Heidegger becomes more ontologically parsimonious and rejects his earlier neo-Aristotelian naturalism. The basic point is methodological: the role of philosophy here is to show how certain concepts biology employs to explain living phenomena are unconsciously drawn from a different source: their original referent lies in the lifeworld of Dasein.[37]

Second, while Heidegger appropriates Uexküll's concept of the organism's *Umwelt* in order to elaborate his conception of what is called, variously, the natural attitude, the lifeworld, or average everydayness, he no longer uses it to underwrite an ontology of life, as he did in the pre-*Being and Time* work. He is careful to defend himself against the charge of biologism—basing ontology on scientific, that is, ontic, data:

> Although this state of Being [the *Umwelt*] is one of which use has been made in biology, especially since K. von Baer, one must not conclude that its philosophical use implies "biologism." For the environment is a structure which even biology as a positive science can never find and can never define, but must presuppose and constantly employ. Yet, even as an a priori condition for the objects which biology takes for its theme, this structure itself can be explained philosophically only if it has been conceived beforehand as a structure of Dasein.[38]

Keep in mind, however, that Heidegger is repurposing the concept of *Umwelt* for his own agenda. The original sense of the concept, in Uexküll's hands, was that all organisms inhabit a subjective world and that nature itself is imbued with meaning, Dasein or no Dasein. While Heidegger holds Uexküll in special esteem among biologists, he thinks Uexküll differs little from them on what matters most: methodological clarity. Heidegger endeavors to "destroy" the concept of *Umwelt* by tracing it to its origin in the human lifeworld and rein in Uexküll's supposedly speculative postulation of animal worlds. Put differently, in *Being and Time* Heidegger begins to differentiate *Umwelt* and world more clearly than in the earlier lecture courses, and he will continue to do so in *FCM*.

Third, Heidegger takes aim at *Lebensphilosophie* and Husserlian phenomenology for failing to determine the being of life and consciousness, respectively. Again, this represents a departure from the pre-*Being and Time* phase because he no longer regards life as a viable philosophical concept; instead, he thinks it is too hazy to serve as a fundamental concept.

Heidegger's relation to *Lebensphilosophie* is ambiguous. On the one hand, he sympathized with his mentor Rickert's contempt for its irrational exuberance. As Krell notes, "Rickert spares none of the enthusiasts of life-philosophy: Schelling, Scheler, Simmel, Dilthey, Bergson, Nietzsche, Spengler, William James, and even Husserl are tainted with it and are accordingly excoriated; all have surrendered rigorously defined concepts and principles for the sake of 'the intuitive.' "[39] On the other hand, Rickert drew a clean line between living and knowing, one Heidegger found too neat. While he agrees that the concept of life is fuzzy and an inadequate ground for philosophy, he thinks Rickert's approach is somewhat of a cop-out that itself relies on an inadequate account of knowing. As Krell notes, "Heidegger repudiates [Rickert's] complacent, not to say smug, separation of living from knowing.... In Rickert, cognition and the concept are 'sheer ghosts,' says Heidegger; and Rickert's philosophy of values and *Weltanschuang* is as vapid as his anemic life."[40] As I explain in the next chapter, in *FCM* Heidegger attributes the nihilism of the present to the crisis between life and spirit, and he thinks that extant views side with either pole but do not question their meaning. For Heidegger, Husserl's main problem is his failure to define the being of consciousness or intentionality. In this context, his critique has to do with the relationship between spirit and life or spirit and nature. In *Ideas II* Husserl investigates material, animal, and spiritual regions, but what bothers Heidegger, Krell observes, is Husserl's attachment to "the relativity of nature and the absoluteness of spirit."[41] Krell correctly points out what this means for Heidegger's view of life: "Heidegger's critique of transcendental phenomenology is a critique of 'spirit.' His formulation of the neglect of the question of being in phenomenology thus owes a great deal to his preoccupation with something other than a philosophy of spirit, something more akin to life-philosophy."[42] So Heidegger sees his project as dispelling a perfect storm of crucial concepts—life, knowing, and spirit—churning confusedly through *Lebensphilosophie*, Neo-Kantianism, and phenomenology. Instead of a philosophy of life, a Neo-Kantian epistemology, or a phenomenology of spirit, Heidegger tries to ground or destroy these concepts, and that is why he assiduously avoids them in the text. This reflects his method of "bracketing," of suspending the use of familiar concepts in order to get at their hidden sense.

Fourth, Heidegger hints that fundamental ontology can ground a "positive" interpretation of life, which is a reversal of his pre-*Being and Time* position. This brings us back to the first of the four points discussed above. Recall that Heidegger's key statement on the place of life in his fundamental ontology, namely, that biology is methodologically founded on the latter, had a corollary: "even if not entirely." Life is tagged with a promissory note: after the groundwork of ontology is laid, then a positive, rather than a privative, account of life can begin. This seems to imply that while life is after Dasein in the order of knowing, it is prior in the order of being. But as Krell points out, the problem runs deeper: "An ontology of life can be neither prior nor posterior to fundamental ontology."[43] It cannot be prior because of Heidegger's methodological assumptions; yet, Krell wonders, "if life does not allow us to get back behind it, is it not because life, and not Dasein, is the ground of ontological analysis?"[44] This seems to be Heidegger's earlier position: an ontology of life is prior to an ontology of Dasein. And it cannot be posterior to fundamental ontology because, if it turns out that life is grounded on something else, such as *physis*, then fundamental ontology is not, in fact, fundamental. Life, then, is left hanging, and if Heidegger does not ground it in *FCM*, then he grounds it nowhere. I return to this issue in the next chapter.

V. The Three Senses of Nature

The lion's share of Heidegger's treatment of nature in *Being and Time* is found in the third chapter of division 1. The two scholars that have provided the most astute analyses of this topic, Hubert Dreyfus and Bruce Foltz, differ slightly on how many senses of nature Heidegger includes: while Dreyfus discerns four, Foltz finds three. In the present section, I sketch Foltz's analysis and then in the following section I discuss his disagreement with Dreyfus in order to pinpoint the problems in Heidegger's account with which they are grappling.

Foltz provides a useful scheme for what he sees as the three senses of nature operative in the existential analytic of *Being and Time*: Productive/Environing Nature, Objective Nature, and Primordial/Poetic Nature.[45] The first sense of nature (productive or environing nature) corresponds to the third sense of world, the environment of our everyday dealings. When we are absorbed within the perspective of everydayness, nature has a meaning only as it bears upon our own affairs, only in so far as it is useful for some project or other. This holds whether we are searching a forest for firewood or checking the clock at the office, whether we understand ourselves to be "in nature" or "in society." Indeed, Heidegger's phenomenology aims to disrupt this commonsense dichotomy between the natural and the social/cultural. As he puts it,

> Along with the public world, the environing Nature is discovered and is accessible to everyone. In roads, streets, bridges, buildings, our concern discovers Nature as having some *definite direction*. . . . In a clock, account is taken of some definite constellation in the world-system. When we look at the clock, we tacitly make use of the "sun's position," in accordance with which the measurement of time gets regulated in the official astronomical manner. When we make use of the clock-equipment, which is proximally and inconspicuously ready-to-hand, the environing Nature is ready to hand along with it.[46]

The disclosure of things as *pragmata*—things of use—and of nature as oriented, however, should not be confused with the mere artifactual. Presumably, the tribesman in the jungle discovers nature in this way, too, despite the absence of manmade structures and sophisticated technologies. What is at issue here is how nature is intended, not how it physically appears, that is, as "natural" or "manmade." To bear a sense, nature must first be intended, but the primary mode of intentionality and understanding has to do with our pretheoretical, everyday concern with our own projects and purposes. The main point is that in the third, environmental sense of world, natural entities are immediately assigned a place within a "matrix of meaningfulness"—the proximate environment—which has a place within the ultimate matrix of the world, which is presupposed but unthematized. We immediately encounter nature as fitting or not fitting within a totality of equipment ordered to an end.

Yet even this is somewhat misleading, because the upshot of the being of the ready-to-hand is that we do not *directly* encounter it. I do not encounter the space key on the keyboard as an individual key fitting into a system of tools while I am typing, but I do nevertheless encounter it *in some way*. When I chop firewood, the wood of the axe handle withdraws into the axe, and the axe itself withdraws into the activity of chopping. This, in turn, is referred to the procurement of firewood, which is referred to the heating of my home, and so on. The point is that the beings involved in this *Gestalt* are what they are insofar as they withdraw before or give themselves to the purpose of the task at hand. Whether the entity in question is artifactual or "natural," in the environment we encounter it *as* produced—as made for a purpose, whether by humans or by nature itself—and as ready to be accessed and manipulated for the sake of an end.

It is important to point out that Heidegger is not saying that productive nature is a mere "projection" of human interests and concerns onto "mere nature," understood as the value-neutral realm of factual, purely given things. He claims that readiness-to-hand is the being "in itself" of productive nature, in the sense that it is just the normal way that nature shows itself to us: "[T]his characteristic is not to be understood as merely a way of taking [these entities], as if we were talking such 'aspects' into the 'entities' which we proximally encounter, or as if some world-stuff which is proximally present-at-hand in itself were 'given subjective colouring' in this way."[47] Productive nature is the way nature shows up proximally and for the most part within our everyday being-in-the-world and corresponds to a perspective prior to the theoretical separation of subject and object.

The second sense of nature (objective nature) corresponds to the idea of nature operative in modern science, a realm of value-neutral, physical objects arranged in various positions in space-time and part of a system governed by causal laws. This sense of nature also corresponds to the concept of nature in traditional metaphysics from which the modern scientific perspective derives. The connection between the two is the notion of being as presence-at-hand in the first and second senses of world. As Foltz points out, "*Vorhandenheit*—the quality of being *vorhanden*, on hand, present at hand—is for Heidegger the phenomenon underlying the interpretation of entities in terms of 'reality,' 'actuality,' substantiality,' and, in fact, all the traditional metaphysical determinations of the being of entities."[48] Indeed, when Heidegger uses the term "nature" in *Being and Time*, it usually carries this second sense, since his chief aim is to show that the basic metaphysical concepts that have been handed down to us—and the sense they bestow to nature—are phenomenologically unfounded. Objective nature can only be encountered when productive nature breaks down, that is, once the circumspection of concern and absorption in one's involvements is disrupted and gives way to the pure, theoretical inspection of entities.

Yet once this theoretical stance arises, it tends to lose sight of the environment and interpret entities as purely present-at-hand. It tends to lose sight, in other words, of the structure of being-in-the-world. Foltz astutely observes that Heidegger's "interpretation of nature as *Vorhandenheit*, and his critique of nature as obscuring our understanding of both ourselves and the world, are in fact an interpretation and a critique of a metaphysical concept of nature rather than a disparagement of the phenomenon itself."[49] The reason for this, again, is methodological: extant understandings of nature must be bracketed so that the full phenomenon can be allowed to emerge.

Heidegger's claim to offer a merely privative rather than positive account of nature in *Being and Time* squares with this as well. His main task, recall, is to offer an ontology of human Dasein, not of nature "in itself."

Indeed, he thinks that the main problem with ontology is that it has been dominated by a conception of time stemming from Aristotle's philosophy of nature. As he explains toward the end of the second division, "In the 'physics' of Aristotle—that is, in the context of an ontology of Nature— the ordinary way of understanding time has received its first thematically detailed traditional interpretation. 'Time,' 'location,' and 'movement' stand together."[50] The virtue of Heidegger's account is to show not only that being and human being were misinterpreted through the prism of nature, but that nature itself was misunderstood through the prism of presence-at-hand. The environing sense of nature was disregarded, and objective nature was granted ontological primacy. Foltz argues that *Being and Time* should not be seen as one-sidedly focused on Dasein to the detriment of nature: "[I] t is [Heidegger's] persistent attempt in the early writings to overcome this decisive 'bias of ancient ontology' (the bias toward understanding entities in terms of the being of a neutrally conceived 'nature') that creates the illusion that Heidegger is somehow 'critical' of nature as such."[51] Heidegger's discussion of nature in *Being and Time* is dominated by the objective sense of nature because he aims at the destruction of the concept, not its validation. This is not to say that the objective sense is somehow an illusion or a wholesale distortion of nature. Though science poses questions to nature that only admit narrow answers, it "nevertheless allows nature to be heard."[52] Objectivity or presence-at-hand is one of the ways that nature shows up, but it is by no means the only or even the primary way.

These first two senses of nature, then, correspond to the ways we habitually intend things in the world (pragmatically and theoretically) and the ways they show up, that is, their modes of being (as ready-to-hand and present-at-hand), but they should not be taken as ontologically definitive. The third sense of nature, according to Foltz, is not so easy to pin down in *Being and Time* because Heidegger's references to it are vague and sparse. Heidegger refers to it when he says that when we disclose nature objectively, "the Nature which 'stirs and strives,' which assails us and enthralls us as landscape, remains hidden. The botanist's plants are not the flowers of the hedgerow; the 'source' which the geographer establishes for a river is not the 'springhead in the dale.'"[53] He also makes a cryptic reference to the concept of nature "in romanticism."[54] This nature is not the fodder for our projects, the object of our investigations, or even the stage for our moral strivings. It finds expression only through poetry, artwork, and contemplation.

However, it would be a mistake to categorize the third sense of nature as merely aesthetic, rather than practical or scientific. Foltz explains how this sense of nature figures in Heidegger's greater and later metaphysical agenda:

"[B]efore it can be set up as an object . . . nature is always already emerging on its own. This self-emergence, according to Heidegger, is what the early Greeks meant by *physis*. . . . It is that sense in which the being of an entity unfolds and emerges from itself while continually returning to itself; both the blossoming of a tree and the beauty of an artwork."[55] While none of this is spelled out in *Being and Time*, Foltz points out, correctly I think, that it is coordinate with that text's project of destroying the history of ontology. In a way, the third sense of nature should not be seen *as a sense* because meaning and significance occur only within the bounds of the world, and this sense of nature ultimately refers to an ontological principle that grounds the world and so cannot show up within it. At stake is the whole traditional opposition between nature and being, "physics" and metaphysics. As Foltz explains,

> Heidegger . . . argued that from the beginning, even before the translation of *physis* by the Latin *natura*, the concept of nature as such has been so thoroughly embedded within the metaphysical tradition that it is hopelessly bound up with the concept of presence at hand. . . . Heidegger's critiques of 'nature' as . . . objectivity and as presence at hand are themselves part of . . . a 'deconstructive analysis' of the concept of nature whose positive terminus is his interpretation of what is entailed by the Greek comprehension of physis as self-unfolding emergence.[56]

The third sense of nature, then, ultimately refers in later works to the being-process itself.

VI. Problems in Heidegger's Account of Nature in *Being and Time*

Here I want to address a cluster of problems issuing from Heidegger's account in *Being and Time* that will set the stage for his later thinking about nature. These issue from Dreyfus's and Foltz's interpretations. Dreyfus's analysis of Heidegger's account of nature in *Being and Time* is similar to Foltz's. He lists four senses of nature. First, he, too, notes that for Heidegger nature normally show up as ready-to-hand or "available," and this availability takes three forms: "natural materials," such as iron and wood; "natural regularities," such as the rising and setting of the sun; and "nature taken up into history," such as a battle*field* or a *country*side.[57] Second, within the sphere of everyday concern, nature also shows up as unavailable, as resistant to our work and world-making activities. The first and second forms (nature as available or unavailable) correspond to Foltz's productive/environing nature. Third, nature shows up as present-at-hand or merely "occurrent" and can be represented in scientific theories. This corresponds to Foltz's "objective nature." Fourth, there is the nature of "primitive peoples and the Romantic poets," which corresponds to Foltz's primordial/poetic nature.[58]

Despite these similarities with Foltz, Dreyfus is concerned with different questions. Whereas Foltz is bent on tracing the development of primordial nature through Heidegger's later writings and advancing it as a sounder basis for an environmental ethic than scientific naturalism or ecocentrism, Dreyfus focuses on the following questions: "(1) Can Heidegger achieve his fundamental ontology, demonstrating that all modes of being, even the being of nature, can be made intelligible only in terms of Dasein's being, and not vice versa? (2) Can he still leave a place for ontic, causal, scientific explanation?"[59] Dreyfus is concerned about the relation of ontological priority between readiness-to-hand and presence-at-hand. If the existence of the present-at-hand is founded on Dasein's being, the legitimacy of scientific knowledge claims appears to be called into question. While there is no doubt that Heidegger means to criticize scientific naturalism (or at least its ontological pretensions), part of his project is to reinsert human beings in the world and overcome the subject/object duality. In this regard, Dreyfus's question bears upon questions in environmental philosophy: like Heidegger, many environmental thinkers criticize the modern conception of nature and seek to resituate human beings in the natural world. However, Heidegger's notion of world in *Being and Time* seems decidedly anthropocentric, and many environmental thinkers are convinced that resituating human beings in nature entails seeing them as ontologically continuous with nature and as shaped by natural processes such as evolution. Heidegger, however, draws a sharp line between the ontological and the ontic and the intentional and the causal orders.

Let us look more closely at the problem Dreyfus is probing. On the one hand, he thinks, Heidegger does not want to reduce the objects of science to the projections and interests of human beings: "[W]hen theory decontextualizes, it does not *construct* the [present-at-hand], but . . . it reveals the [present-at-hand] which was already there in the [ready-to-hand]."[60] Dreyfus reiterates this view when he explains Heidegger's notion of occurrent nature: "Scientific observation can thus reveal a universe unrelated to human for-the-sake-of-which's. This is the nature whose causal powers underlie equipment and even Dasein itself insofar as it has a body."[61] Hence, it appears that the present-at-hand must be prior.

On the other hand, Dreyfus notes that Heidegger seems to maintain that the present-at-hand is founded on the ready-to-hand. The inspection of an entity is described as a "deficient mode of concern" and only arises when concern has been disrupted. And recall that Heidegger is adamant that when we take things as ready-to-hand, that is, not just an interpretation we project onto the things.[62] This appears to contradict the view that Dreyfus imputes to Heidegger. In describing environing nature, Heidegger claims that *"readiness to hand is the way in which entities as they are 'in themselves' are defined ontologico-categorially."*[63] The trick is that Heidegger argues there is an "objective" or "real" aspect of the ready-to-hand and that, conversely, there is a "subjective" or "ideal" aspect to the present-at-hand, which we usually take to be objective and mind-independent, but is actually encountered only through a modification of everyday concern. Here Heidegger shows us he is aware of the paradox: "Yet only by reason of something present-at-hand, 'is there' anything ready-to-hand. Does it follow, however, granting this thesis for the nonce, that readiness-to-hand is ontologically founded upon presence-at-hand?"[64] Moreover, both of these modes of being are founded on being-in-the-world. As he points out, "In Interpreting [present-at-hand and ready-to-hand] entities within-the-world . . . we have always 'presupposed' the world. Even if we join them together, we still do not get anything like the world as their sum."[65] The point is that it does not seem possible to answer both of Dreyfus's above-mentioned questions in the affirmative: if all modes of being can be traced back to Dasein's being-in-the-world, then there seems to be no objective foundation for scientific explanations (question two); yet if there is such an objective foundation, then fundamental ontology cannot claim to ground all modes of being (question one). If the latter, then we are thrown back upon the Cartesian problems of interaction that fundamental ontology is intended to disarm: namely, how intentionality (consciousness) and causality (nature) are related. Heidegger does not account for how the spheres of intentionality and causality relate to one another. If he maintains that while there can be no being or truth without Dasein, yet there can be beings without Dasein, then how are we to describe them if meaning and significance obtain only within the world? And if we are to maintain that they are indeed things, that there is a cosmos subtending and embracing human history, then how can it have no sense for us and why can we not say that it has its own distinctive temporality and way of being? The primordial sense of nature seems to be a kind of noumenal thing-in-itself that we must acknowledge but that we cannot hope to comprehend. And this conception of nature—which, as I explain in later chapters, closely resembles Kant's "sublime" nature—is where Heidegger ultimately ends up.

Dreyfus thinks that Heidegger has a solution to this dilemma: he calls this Heidegger's "hermeneutic realism." According to this view, Heidegger "demonstrate[s] that although natural science can tell us the truth about the causal powers of nature, it does not have a special access to ultimate reality."[66] In a sense, Dreyfus is merely arguing that Heidegger attacks scientism in order to save science. This position is "hermeneutic" because, similar to Thomas Kuhn, it recognizes that science does not happen in a vacuum: its practitioners are always operating in a social context and a particular tradition laden with presuppositions stipulating what the fundamental problems are and which data are significant. Scientists cannot escape the sociohistorical contexts in which they move and have their science. As Dreyfus explains, "science cannot justify a metaphysical realism claiming to have an *independent argument* that nature has the structure science finds and that science is converging on the one true account of this independent reality."[67] For Heidegger, this is a consequence of fundamental ontology: the basic mode of Dasein's understanding is interpretive; in philosophy of science parlance, all observation is "theory-laden," and there is no "view from nowhere." Now, this is not to say that the discoveries of science are purely cultural constructions or are ultimately referred back to pragmatic human concerns; Heidegger is not, Dreyfus insists, a scientific instrumentalist. Thus this position is "realist" in the sense that nature and the natural objects studied by natural science do exist apart from human beings, but science cannot overstep its bounds and assert that the sense and structure it attributes to these beings reveals what they are "in themselves."

The broader claim that underwrites this view is Heidegger's confusing contention that there can be *beings* without Dasein, but not *being*. As he puts it, "Entities *are*, quite independently of the experience by which they are disclosed, the acquaintance in which they are discovered, and the grasping in which their nature is ascertained. . . . Being 'is' only in the understanding of those entities to whose being something like an understanding of being belongs."[68] Dreyfus notes what this spells for nature: "It seems that while natural entities are independent of us, *the being of nature* depends upon us."[69] This claim is not made without warrant, but we should notice that it contradicts Foltz's interpretation above: "[I]t does not even belong to the being of nature as such to be within the world."[70] If the being of nature depends on us, then there can be no being of extra-worldly nature, and we are stuck in a kind of constructivism. Yet Foltz is surely on to something in thinking that Heidegger does have an alternative, extra-worldly sense of nature in mind because Heidegger clearly states that the being of nature does not show up in the world. How can this issue be resolved?

For a being to be encountered *as* a being, that is, to have a sense, it must be implicated in a totality of significance, which is to say that it must be annexed into the structure of worldhood. Yet it can only be so encountered on the basis of Dasein, the sole entity, Heidegger claims, whose being is an issue for it, who is ontological. Dreyfus clarifies this point:

> Only Dasein can make sense of things. So the intelligibility of each domain of things, or the understanding of the way of being of each, including that of natural things, depends upon Dasein. But nature as a being, or as a set of beings, does not depend on us, for one way Dasein can make sense of things—find them intelligible—is as occurrent, i.e., as not related to our everyday practices.[71]

Now, as we have seen, the two primary modes of intraworldly being are the ready-to-hand and the present-at-hand, which Dreyfus calls the available and the occurrent. So neither of these senses or modes of being can legitimately be attributed to an entity falling outside the ambit of worldhood. However, it seems to me that this is exactly what Dreyfus demands with hermeneutic realism: specifically, that entities are present-at-hand or occurrent independently of Dasein, and that this is what Dasein is discovering in scientific investigation. He sums up his view here: "Heidegger thus holds a subtle and plausible position beyond metaphysical realism and antirealism. *Nature* is whatever it is and has whatever causal properties it has independently of us."[72] Causality is said to obtain in the natural "world" quite independently of Dasein's understanding activity, and there is an ontic structure, a logos, to the universe, though Dasein, on account of its worldly constitution, can never quite draw a perfect picture of it. This view is reflected in Dreyfus's characterization of the forms of nature discussed above, where the ontic, causal properties of things constrain the ways they can be encountered as ready-to-hand. But since this realism is hermeneutic, it means that the causal properties we ascribe to natural entities are not their ultimate nature, and since Heidegger is committed to the claim that intelligibility is founded on Dasein, then natural entities cannot be intelligible in themselves.

I concur with Foltz that Dreyfus goes too far in claiming that Heidegger espouses an instrumentalist view of nature in *Being and Time*.[73] Dreyfus points to the following quotation to buttress this claim: "The wood is a forest of timber, the mountain a quarry of rock; the river is water-power, the wind is wind 'in the sails.' "[74] Dreyfus's hermeneutic realism seems to contradict this view: he says repeatedly that the ready-to-hand is rooted in the present-at-hand, and that this mode of being depends on an extraworldly kind of being of which Dasein cannot speak. Dreyfus has also described his view as "robust realism," which holds that "science can in principle give us access to the functional components of the universe [i.e., nature] as they are in themselves in distinction from how they appear to us on the basis of our daily concerns, our sensory capacities, and even our way of making things intelligible."[75] Clearly, then, on Dreyfus' own analysis Heidegger cannot be said to subscribe to an instrumental sense of nature. It seems that the quotation Dreyfus seizes upon is merely a case of Heidegger giving a *description* of how nature shows up from the standpoint of everyday concern, not making an *ontological claim* about nature.

However, I think Foltz fails to appreciate the gravity of the problem with which Dreyfus grapples. While his own extensive analysis of primordial nature in Heidegger's later writings fills the lacuna left by Dreyfus's account, I do not think it deals sufficiently with the relationship between intra- and extraworldly nature and the problem of fitting human beings into an ontological continuum with natural beings. And the key to all of this is that neither Dreyfus nor Foltz reckons with Heidegger's investigations into the philosophy of biology: when they are talking about "science," I think they are referring primarily to physics. I will return to this issue in the next chapter.

Dreyfus concedes that on Heidegger's view one-to-one *correspondence* between theory and reality is not possible, but insists instead that *convergence* of theory with reality is. That is fine, as far as it goes. But it still begs the question of whether and to what extent material causality affects, conditions, or underlies intentionality. Moreover, shifting the paradigm to convergence rather than correspondence seems to imply that though present science's explanation of the workings of nature only discovers an aspect of nature "in itself," in the long run its theories will provide a fuller, more comprehensive explanation, yet one still in terms of presence-at-hand or mere occurrence.

This qualified scientific realism seems inconsistent with Dreyfus's claim several pages later that "Heidegger made it clear that the 'evidence' that there is a nature independent of us is provided not by natural science but by anxiety. . . . Dasein is presumably thrown into nature, but the nature Dasein is thrown into need not be thought of as the unstructured, viscous being-in-itself as in Sartre. Anxiety reveals nature as pure otherness, but this does not imply that nature has no ontic structure."[76] Heidegger is clear that anxiety is not to be thought of as a psychological phenomenon, a subjective mood that colors what is, objectively and in itself, a present-at-hand, value-neutral world, but rather as an ontologically disclosive phenomenon that shows us something important (and something more fundamental) about the world. But if the cosmos does indeed have an ontic *structure*, then why wouldn't humans be able to encounter it and make meaningful statements about it? How can the extraworldly or primordial nature be both radically other and alien—as Otto Pöggeler puts it—and yet still possess a structure, not be a Sartrean "viscious being-in-itself"? What is the proper ontological register for discoursing about extraworldly beings? In Heidegger's later work, the experience of anxiety will be reconceived in terms of aesthetic and poetic—some might say mystical—experience, and this will be intimately connected with the experience of nature as *physis*, which in turn comes to be more or less synonymous with being itself. Moreover, this experience becomes less about the self's confrontation with its own nullity and more about the natural manifestation of things. As Graham Parkes puts it, "The work of art, whose essential nature cannot be appreciated if it is taken as an implement or an object of scientific investigation, is to be seen here *as a paradigm of things in general.*"[77] The perspective called authenticity in *Being and Time* is later interpreted along the lines of aesthetic and poetic experience, and it ceases framing objects as tools and instead encounters them as manifestations of *physis*. Indeed, as we will see, in the 1930s Heidegger becomes convinced, largely through the influence of Nietzsche, that the solution to the problem of nihilism involves adopting a more aesthetic comportment to the world.

In this later discourse, human beings are no longer portrayed as anxious aliens within nature, but as native inhabitants of the earth, as being at home in nature. Even so, this discourse is thoroughly poetic and vague rather than rational or phenomenological and does not articulate a picture of an evolving, hierarchical cosmos reminiscent of the great chain of being. This is why Jonas criticized Heidegger. Zimmerman pinpoints the heart of Jonas's critique:

> Jonas concluded that what Heidegger really objected to was placing humans in *any natural scale*. Though condemning the technological domination of nature, Heidegger was never a "biocentrist," [as some deep ecologists have claimed] but rather a Gnostic, who viewed humans as aliens adrift in an indifferent or even hostile cosmos.[78] Jonas's alternative is an ethics grounded on the "objective assignment by the nature of things. . . . Only

an ethics which is grounded in the breadth of being, not merely in the singularity and oddness of man, can have significance in the scheme of things."[79]

Heidegger occasionally invokes the term "cosmos" but neglects to explain how he understands its meaning throughout the philosophical tradition and fails to venture how it might be reinterpreted in a phenomenologically palatable way. He says, for instance, that "[t]he cosmos can be without human beings inhabiting the earth, and the cosmos was long before human beings ever existed."[80] This squares with the thesis that there can be beings, though not being, without Dasein. But it appears to conflict with his assertions about nature and time elsewhere. In the *Basic Problems of Phenomenology*, for instance, he claims, "There is no nature-time, since all time belongs essentially to Dasein."[81] Similarly, in *Being and Time* he says, "Even Nature is historical. It is not historical, to be sure, in so far as we speak of 'natural history'; but Nature is historical as a countryside, as an area that has been colonized or exploited, as a battlefield, or as the site of a cult."[82] Human temporality is never grafted onto or integrated with natural or cosmic time, and human being is never quite squared with natural or cosmic being. This is what motivates the charge of anthropocentrism.

In a lecture course from 1925, Heidegger makes a cryptic remark that bears upon this issue: "The question of the extent to which one might conceive the interpretation of Dasein as temporality in a universal-ontological way is a question which I am myself not able to decide—one which is still completely unclear to me."[83] In other words, what if Heidegger abandoned the ontic priority of the question of being and universalized the ontological structure of being-in-the-world? What if all beings, all the way down from apes to amoebas to atoms, were shot through with some kind of interiority or transcendence? What if, after successfully critiquing the substance metaphysics that had pervaded and perverted the philosophical tradition, Heidegger had, like Whitehead with his notion of process and the method of descriptive generalization, forged a new conceptual scheme for talking about the being of beings in the natural cosmos?[84] If something like this were possible, then there would be a transition from fundamental ontology to cosmology, primordial nature would no longer be conceived of as pure otherness, and there would no longer be the ontological dichotomy of human existentiality and natural categoriality. Moreover, this may have provided the basis for what Heidegger will call the "inner unity of science and metaphysics." By granting temporality to nonhuman entities and nature itself, Heidegger would not run into the paradox of saying that, on the one hand, "all time belongs essentially to Dasein," and, on the other, "the cosmos was long before human beings ever existed."[85] Moreover, it would lay the groundwork for situating human beings within a greater cosmological, evolutionary vision. As I explain in the next chapter, the connecting link here is a reinterpretation of the concept of life.

Conclusion

Heidegger's early account of nature appears to leave us with a choice between viewing nature as present-at-hand (composed of entities with a discernible ontic structure discoverable by natural science) or as a pure otherness that can only be expressed aesthetically. I want to suggest that this is where *Being and Time* leaves us with regard to the concept of nature, and that this *aporia* gives rise to two incompatible ways of thinking about nature. One way holds, with questionable consistency, that the *being* of present-at-hand entities is ontologically dependent on Dasein and that present-at-hand entities exist independently of Dasein but are discoverable by the natural sciences, and *precisely as* independent of Dasein (Dreyfus' hermeneutic or robust realism). This position is potentially inconsistent because it begs the question of the being of these world- and Dasein-independent entities, but it does broach the important question of whether philosophy can furnish a more adequate approach to these entities than scientific naturalism. The other way treats nature as a radical alterity, an altogether strange and incomprehensible kind of being that can only be described metaphorically, aesthetically, and poetically. This way interprets nature somewhat like Kant's notion of the sublime in the third critique.

Hence two roads lead out from *Being and Time*. One leads to the fundaments of Heidegger's later philosophy: his turn toward the history of being, his enigmatic notion of earth, his retrieval of *physis* as primordial nature, and his critiques of humanism and modern technology. The other road, already hinted at in his early Aristotle lecture courses, leads to an investigation of theoretical biology and ecology in *FCM*, in which Heidegger approaches the being of humans, animals, and inanimate beings such as stones analogically through the concept of world. The distinctive part of this analysis is that it is the sole place where Heidegger entertains a continuum view of humanity and nature and tries to "plug" Dasein back into the ontic order. It is, in other words, his closest attempt to escape the limitations of transcendental phenomenology and fundamental ontology and articulate a nonreductive naturalism: a view in which human beings are ontologically continuous with nature and natural processes such as evolution, yet retain their distinct intentional and interpretive powers. In the next two chapters, I want to trace these two roads, show their incompatibility, and argue that Heidegger would have done well to take the second road.

4

Back to Life

Organism, Animal, and Umwelt *in* Fundamental Concepts of Metaphysics

Though Heidegger does challenge the domination of mechanistic materialism, link it to the phenomena of nihilism and anthropocentrism, and advance an alternative view of nature, a robust account of life is a serious lacuna in his oeuvre. As David Farrell Krell notes, "However much Heidegger inveighs against life philosophy his own [early] fundamental ontology and [later] poetics of being thrust him back onto [it] again and again."[1] Heidegger's difficult position on animal life has received much attention in recent scholarship. However, as Frank Schalow has recently pointed out, one of the "glaring omissions" in this scholarship is any regard for the place of evolutionary theory in Heidegger's account of not only animals, but human beings and nature at large.[2] I am going to extend Schalow's observation to show that one of the deepest problems in Heidegger's view of life is his almost utter neglect of the question of evolution. Concerned that any capitulation to evolutionary theory would amount to a kind of biologism and a vindication of mechanistic materialism, Heidegger treated evolutionary theory as anathema. This is the same pattern we find in his position on values. In both cases, Heidegger excludes important concepts from his positive account of nature by assuming that *narrow* interpretations of them are the *only* interpretations possible: that evolution is necessarily reductionist and that values are inherently anthropocentric and blind to Being. I submit, however, that some integration of evolutionary and value theories is a necessary condition for an environmental ethic. Despite his critiques of modern forms of anthropocentrism and the Being-centric cast of his later thought, Heidegger's exclusion of these concepts renders his philosophy anthropocentric by default—the sole imperative Being/*physis* offers us is to "let beings be," a notion that may have a "holistic" ring, but that offers scant practical guidance in actual environmental issues and cannot orient or place us in nature.

All of this is to say that, contra Foltz, Jonas and Löwith were basically correct: by refusing to place human beings in any natural scale, he cannot give us an ethics rooted in the cosmos, in nature as such. This bears upon the contemporary project of naturalizing phenomenology. As Iain Thompson explains, "Eco-phenomenology's guiding idea, put simply, is that uprooting and replacing some of our deeply-entrenched but environmentally-destructive ethical and metaphysical presuppositions can help us heal the earth."[3] This project has two fronts, one ontological and the other ethical: "[A]ll phenomenological approaches seek to undermine the mind/world divide,"[4] and "[they] are all committed to some type of ethical realism."[5] I do not purport to answer the question of whether and to what extent phenomenology can be "naturalized" here. In my view, two questions have to be asked first: Whose phenomenology? Which naturalism? I will say, however, that in the case of Heidegger's phenomenology, while it does offer useful criticisms of scientific naturalism, it does not offer us a positive vision of humanity's place in nature that motivates a value-based ethic and integrates modern science. Here I appeal to Iain Thompson's useful, if contentious, distinction between two kinds of ethical realism: naturalistic and transcendental.[6] As Thompson explains, "Nietzscheans and Husserlians gravitate toward a naturalistic ethical realism, in which 'good' and 'bad' are ultimately matters of fact, and our 'values' should be grounded in and reflect these proto-ethical facts."[7] On the other hand, Heideggerians and Levinasians, he says, "articulate a transcendental ethical realism, according to which we can indeed discover what really matters when we are appropriately open to the environment, but what we thereby discover is neither a 'fact' nor a 'value' but rather a transcendental source of meaning that cannot be reduced to facts, values, or entities of any kind."[8] The latter is exactly Heidegger's "third sense of nature" as *physis*, which he also refers to as the "law of the earth." In this sense, while Heidegger's early views are more in line with value realism, his mature view can in no meaningful way be "naturalized." It also pinpoints the way in which his view is anthropocentric. As Thompson notes, transcendental ethical realism "results in a more humanistic perfectionism which emphasizes the cultivation of distinctive traits of Dasein." But it is not clear why the "fact" that nature is a "source of meaning" should lead us to respect or preserve it. If it is neither intelligible nor valuable, why is there any reason for a particular ethical stance? Couldn't ethical paralysis or neglect be just as legitimate a response to such a mysterious nature? In other words, this transcendental position does not seem able to overcome the problem of nihilism.

I proceed as follows. In the first section, I return to the theme of nihilism discussed in the previous chapter. Heidegger's focus on nothingness increases after *Being in Time* in the 1929 essay "What Is Metaphysics?" and is central to the first part of *FCM*, in which Heidegger analyzes boredom, a form of nihilism, as a fundamental attunement of the present age.[9] In the latter text, Heidegger stresses the importance of retrieving the Greek concept of *physis* in order to overcome nihilism. This is the context in which his investigations into life, biology, and animality take place: to combat nihilism and recover a more adequate sense of nature.

In the second section, I trace Heidegger's treatment of life, biology, and animality in *Fundamental Concepts of Metaphysics*. My position is that Heidegger's comparative investigation in this text was actually on the right track, in part because it entertained a hierarchical view of humanity's place in nature. Moreover, it actually signaled a return to his pre-*Being and Time* investigations into Aristotle's writings on life and animals. These aspects of his work in the 1920s constitute a strain of naturalism much like that developed by those in the tradition of biophenomenology, such as Jonas and Grene, and sought by contemporary eco-phenomenologists. In my view, the moves toward the transcendental approach in *Being and Time* and toward poeticizing in Heidegger's later thought constitute a wrong turn in the philosophy of nature, and that the pivot point is the concept of life. In his own monograph on Heidegger's life philosophy, Krell provocatively suggests that a proper study of life "would have to proceed to the 'new interpretation of sensuousness' that Heidegger promised at the end of his first lecture course on Nietzsche, promised but never delivered."[10] As I will show in the final chapters, Heidegger would have done well to follow the Nietzschean leads in his thought path.

In the third section, I review contemporary debates over Heidegger's position on animals, drawing on the interpretations of authors such as Will McNeill, Jacques Derrida, and Giorgio Agamben. I isolate three central issues surrounding these debates: 1) evolution, 2) hierarchy, and 3) the human/animal divide. I argue that both Heidegger's and many of his interpreters' position on the latter are compromised by their failure to engage evolutionary theory, their uncritical rejection of hierarchy, and their embrace of what I call alterity theory. Moreover, I conclude by showing that we can triangulate Heidegger's position on the ontological status of life by tracing the tension between the Kantian and Aristotelian strains in his work. On the one hand, Heidegger follows Kant in refraining from claiming teleology as a constitutive principle of living being and in eschewing a robust metaphysical biology; as we saw in the previous chapter, the Kantian strain begins to dominate in *Being and Time* (though, again, after the late 1920s, Heidegger disavows some key aspects of Kant's approach). On the other hand, Heidegger sees Aristotle's understanding of motion as the crucial but forgotten breakthrough in ancient ontology, and this understanding not only informed his account of human existence but led him to reject the Darwinian biology that Kant's endorsement of mechanism underwrote and to explore nonreductive approaches to living and animal being. Nevertheless, after 1929–30, Heidegger abandons this neo-Aristotelian biology, and his later philosophy of nature is largely an inheritance from Kant.[11]

I. Nihilism in Heidegger's Late 1920s Writings

Heidegger's focus on the nothing continues in two works from 1929: the essay, "What Is Metaphysics?" and *FCM*. In the former text, Heidegger reiterates his main ideas from *Being and Time* in less technical language and essentially replaces "being" with "the nothing." He begins by challenging the conventional wisdom that the nothing should be thought of as a kind of logical operator or an intellectual act, that is, as negation. This view implies that the nothing is not really worthy of investigation as an aspect of *being*; it is only a functional term relative to inquiries into *beings*. Negation is always a negation *of* beings. The nothing does not register, for instance, on the conceptual radar of science. As Heidegger says, "The nothing—what else can it be for science but an outrage and a phantasm? If science is right, then only one thing is sure: science wishes to know nothing of the nothing."[12] But Heidegger insists that this logical, intellectualized framing of the question of the nothing misses something essential, something that can only be accessed through a "fundamental experience of the nothing."[13] Heidegger describes this experience in terms of anxiety, as in *Being and Time*, but he also emphasizes that the nothing is the condition for the possibility of any encounter with beings at all: "In the clear night of the nothing of anxiety the original openness of beings as such arises: that there are beings—and not nothing. But this 'and not nothing' we add in our talk is not some kind of appended clarification. Rather, it makes possible in advance the revelation of beings in general."[14] He states bluntly that "Dasein means: being held out into the nothing. . . . Without the original revelation of the nothing, no selfhood and no freedom."[15] The original nihilation is not some intellectual operation performed by Dasein, but issues from being itself, as the ontological process of beings coming to presence; the very process that the Heidegger of the 1930s equates more and more with *physis*. This original, ontological nihilism is the seed of metaphysics: "Human existence can relate to beings only if it holds itself out into the nothing. Going beyond beings occurs in the essence of Dasein. But this going beyond is metaphysics itself. This implies that metaphysics belongs to the nature of man. . . . Metaphysics is the basic occurrence of Dasein. It is Dasein itself."[16] As we will see later on, this insight will become the basis of Heidegger's narrative of the history of being, according to which the history of Western metaphysics is none other than the history of nihilism.

In this essay, Heidegger also touches on a theme that will become the centerpiece of his brush with nihilism in *FCM*: profound boredom. As he says,

> no matter how fragmented our everyday existence may appear to be . . . it always deals with beings in a unity of the "whole," if only in a shadowy way. Even and precisely when we are not actually busy with things or ourselves, this "as a whole" overcomes us—for example in genuine boredom. . . . Profound boredom, drifting here and there in the abysses of our existence like a muffling fog, removes all things and human beings and oneself along with them into a remarkable indifference.[17]

In *FCM*, Heidegger searches for a "fundamental attunement" in order to investigate a set of fundamental philosophical concepts. All fundamental concepts, he says, have the following two characteristics: they have 1) a comprehensive scope, a reference to beings as a whole, and 2) a reference to Dasein's relationship to beings as a whole—to its factical life. They can only be accessed through an actual lived experience of beings as a whole, what Heidegger calls an attunement, and what is variously translated as mood or state of mind (*Stimmung*). But he has in mind not a "mere emotional state or event" or some sort of private psychological phenomenon, but a stance toward the world that discloses beings in a certain way.[18] And some attunements, such as anxiety and boredom, are special in that they grant us a wider perspective on our situation as a whole and are thus the condition for the posing of certain philosophical questions.

Heidegger already did something like this in *Being and Time* through his discussion of anxiety, but the investigation of boredom in *FCM* has a slightly different motivation: Heidegger is trying to determine the fundamental attunement of his own historical age, not just that of Dasein proper. As such, he considers four prominent "interpretations of the contemporary situation": Oswald Spengler, Ludwig Klages, Max Scheler, and Leopold Ziegler. These interpretations are all influenced in some way by Spengler's famous theme of the "decline of the West." As Heidegger describes it,

> Reduced to a formula, it is this: the decline of life in and through spirit. What spirit, in particular reason, has formed and created for itself in technology, economy, and world trade, and in the entire reorganization of existence symbolized by the city, is now turned against the soul, against life, overwhelming it and forcing culture into decline and decay.[19]

The "ecological crisis" is a contemporary example of such an interpretation: it is a pithy, diagnostic phrase that captures a collective intimation about the trajectory of the times. While these stories are not quite "idle talk" and "chatter"—Heidegger calls them the "higher journalism" of the times—they nevertheless traffic in "stereotypes" that call for clarification. They are united by their concern for a collectively recognized but vaguely articulated crisis in the relation between spirit and life: "[T]he essential thing that matters to us is the fundamental trait of these interpretations, or better, the perspective in which they all see our contemporary situation. In terms of a stereotype once more, this perspective is that of the relation between life and spirit."[20] All of the interpretations offer some solution to the perceived imbalance of these two forces, but Heidegger thinks that the real problem is that we do not know what these terms—life and spirit—actually *mean*. The phenomenologist must attempt to penetrate these stereotypes of the present in order to clarify the matter they point to and partially conceal.

Now, Heidegger claims that these interpretations are all determined by Nietzsche's dyad of the Dionysian and the Apollonian and that they point to the phenomenon of boredom. As we will see later on, Heidegger is adamant that Nietzsche is the philosopher of the present and of the modern age: he recognizes the fundamental problems and sets up the conceptual horizon in which the contemporary age moves. While I will not go into detail here about the Nietzschean dyad, suffice it to say that he thought these two forces, originally captured in Greek tragedy, had grown out of balance in modernity. More specifically, modernity saw the over-development of the Apollonian power (roughly, of reason, conformity to plan, measure, the conscious, the *logos*, the law) over the Dionysian (again very roughly, the passions, the sensuous, the sublime, the unconscious, the *mythos*, and the mysterious). Put differently, modernity saw the conquest of life by the spirit.

Heidegger sees the popularity and power of these narratives as a symptom of the dominance of the "philosophy of culture," which sees man as a "symbolic animal." The soul or spirit is expressed in symbolic forms "that bear an intrinsic meaning and which, on the basis of this meaning, give a sense to existence as it expresses itself. . . . Man . . . is in this way set out in terms of the expression of his achievements."[21] Two things are worth noting here. First, Heidegger is reacting to an assumed binary between *Natur* and *Geist*. The implication of the philosophy of culture is that human life, insofar as it is "natural," is meaningless; that natural existence itself is meaningless. Meaning is something that needs to be *produced* over and beyond nature, and its production is what we call culture. To study the human essence, then, we should turn to psychology and anthropology. The second point is closely related. Heidegger insists that the philosophy of culture produces a "setting-out" (*Dar-stellung*) of human beings "in terms of the expression of [their] achievements," a *representation* or objectification of life, not their Dasein, their factical, lived existence.[22] This is why Heidegger thinks that the cultural narratives are superficial: they do not "take hold of us or even grip us." They do not reflect our actual experience of the world.[23] They are prefabricated interpretations that relieve us of the burden of encountering our situation as it is, uniquely and "authentically." Heidegger questions why we are susceptible to such narratives and answers that we lack a common destiny. Our existence is impoverished: we hunger for meaning, we desire to be interesting, and we do so, Heidegger suggests, because "we have become bored with ourselves."[24] So Heidegger traces the problem of nihilism here not to the cultural narratives *themselves*, but to the *need* for them.

Keep in mind that this lecture course is largely concerned with further clarifying the phenomenon of world, and that in the second part, Heidegger characterizes the animal as "poor in world." His description of animal existence bears a striking resemblance to that of inauthenticity in *Being and Time*. Nihilism, the rootlessness of the modern age, means the lack of world, of a shared context of meaning and a historical task, *and* the attempt to posit or create such a meaning. It is no accident that here Heidegger couples his account of nihilistic boredom with an investigation of theoretical biology, animality, and the metaphysical status of life. He intimated that nihilism had something to do with humanity's changed relationship to nature, yet after this lecture course, he never pursued the task of working out a framework that integrates humans in the continuum of nature and subjects them to natural processes such as evolution.

II. Life and Animality in
Fundamental Concepts of Metaphysics

As I explained in chapter 2, there are roughly three phases to Heidegger's thinking about animals: 1) pre-*Being and Time*, Heidegger weaves together elements from the biological works of Aristotle and Uexküll into an ontology of life that ascribes disclosedness, being-in-the-world, being-with (*Mitsein*), and the capacity for signification to animals; 2) in *Being and Time* and *FCM*, he claims that they have an *Umwelt* (environment) but, as *FCM* puts it, they are "world-poor," bound by their drives; 3) in *Introduction to Metaphysics* and after, he insists that animals have *neither* world *nor* environment and generally seems to lose interest in the concept of the animal.

FCM is Heidegger's most extensive examination of life and animality, where he revisits the biological themes from his pre-*Being and Time* writings through the lens of his fundamental ontology. I will highlight three major aspects of the text relevant to our discussion: 1) world and the question of essentialist hierarchies, 2) the phenomenology of organism and *Umwelt* as an alternative to mechanism and vitalism, and 3) the distinction between animal behavior and human comportment.

Heidegger's overriding concern in the second part of this course is not life or the essence of animality, but the concept of world. He explains his approach thus: "Man has world. But then what about the other beings which, like man, are also part of the world? Or does the animal too have world, and if so, in what way? In the same way as man, or in some other way? And how would we grasp this otherness?"[25] In *Being and Time,* Dasein was treated as the ontological being *par excellence*, and for methodological reasons all others were relegated to the sphere of the ontic, but here Heidegger insists that the commonsensical distinctions we make among humans, animals, and nonliving things should be taken seriously. As Brett Buchanan notes, "In order to authoritatively state that Dasein is distinctive in its relation to world—that Dasein has a world, while other living and nonliving things may not—would in the end require an ontological analysis of other beings."[26] Heidegger thus begins with three provisional theses: that humans are world-forming, animals are world-poor, and stones are worldless.

In section 46, Heidegger insists that the notion of "poverty" in world "does not entail hierarchical assessment," but I want to insist that it does entail hierarchy in a descriptive sense.[27] He seems to invoke something like Lovejoy's "great chain of being," an important feature of which was the ontological principle of continuity. According to the *Encyclopedia Brittanica*, "The principle of continuity asserts that the universe is composed of an infinite series of forms, each of which shares with its neighbor at least one attribute."[28] Heidegger seems confused on this point. On the one hand, he says that the rich/poor distinction is one "of degree in terms of levels of completeness with respect to the accessibility of beings in each case"—which sounds quite similar to his pre-*Being and Time* embrace of the *De Anima* and generally in line with the principle of continuity—and much of the analysis suggests that the provisional theses are true.[29] On the other hand, he thinks that the "self-evident" nature of the distinction quickly breaks down, that the comparison "allows no evaluative ranking or assessment with respect to completeness or incompleteness,"[30] and that "the term 'world' itself cannot express quantity, sum total, or degree with respect to the accessibility of beings."[31] The problem here is methodological access: "[W]e find ourselves moving in a circle when we presuppose a certain fundamental conception concerning both the essence of life and the way in which it is to be interpreted" and expect this to "lead us to a fundamental conception of life."[32] The problem is the hermeneutic circle; in the case of life, we cannot access its being "in itself" because our presuppositions about what life is issue from the kind of being we are, so we can only define it in terms of what it is not. This is exactly why life was bracketed in *Being and Time* and approached only privatively.

As Heidegger sketches his phenomenology of life, he focuses mainly on animals, since they are the ambiguous middle case, and his analysis aims to pry apart the senses in which animals both do and do not have a world. In doing so, he draws on research in zoology and biology that he sees as spearheading a revolution in the understanding of life. Heidegger points out both the importance and the difficulty of working out a "metaphysical interpretation of life" and alludes to the "inner unity of science and metaphysics."[33] In his interpretation of the organism, he proposes that we see organisms not as present-at-hand things, but in terms of their potentiality. This is similar to his approach to human being, where we should approach the phenomenon in terms of its existence possibilities, not its empirical actuality. The organism should not be divvied up into its organs and their functions and then patched back together as merely the sum of its parts; rather, the organs and parts should be seen as coordinated by its being as a whole. This is what distinguishes the organism from a machine: "Self-production in general, self-regulation and self-renewal are obviously aspects which characterize the organism over against the machine and which also illuminate the peculiar ways in which its capacity and capability . . . are directed."[34] He notes that behavior cannot be explained as immediate reactions, or the conditioned aggregate of immediate reactions, to environmental stimuli: "The regulation which always lies embedded in the capacity as such is thus a structure of instinctually organized anticipatory responses in each case which prescribes the sequence of movements."[35] This relates to the notion of drives, which have a futural aspect by which the animal is always driven away from itself toward its environment, a kind of "drive intentionality." Though the animal is driven beyond itself, it still remains "within" itself. Heidegger calls this phenomenon behavior and captivation: "[B]ehavior is precisely an intrinsic retention and intrinsic absorption, although no reflection is involved."[36] Heidegger's understanding of drives here is ambiguous. Krell, for one, thinks that "Heidegger does not define the life of drives with any originality or penetration: he accepts the common notion of the drive as a tension resulting from a surfeit of energy, a damming up of force," which is to say that he interprets drives mechanistically, which is to say further that his notion of drives sabotages his project.[37] McNeill, however, thinks that Heidegger's animal is "still able to move and 'act' in a manner that is not wholly or exclusively determined by its immediate environment."[38] Though the organism is more than its body, it cannot escape the closed circuit of its drives. Thus the animal does not have genuine transcendence, like humans: it cannot recognize something *as* something. It lacks the capacities for understanding and interpretation. This is why it is world-poor, and why it is "separated from man by an abyss."[39]

Heidegger focuses on two innovations in biology: Hans Driesch's holistic view of the organism and Uexküll's theory of animal *Umwelts*. While Driesch rescued the forest of the whole organism from the trees of its parts, he failed to see that it cannot be treated as an atomic unit in isolation from its environment. For Driesch, "the totality of the organism coincides as it were with the external surface of the animal's body."[40] Heidegger seems to think that this is what leads him, along with other vitalists, to mistakenly posit some force or entelechy internal to the organism that guides its behavior and development.

To avoid both mechanism and vitalism, Heidegger returns to his earlier interest in Uexküll's work to show that the organism must be seen not just as more than the sum of its parts, but as structurally coupled with its particular environment: "The organism is not something independent in its own right which then adapts itself. On the contrary, the organism adapts a particular environment *into* it in each case."[41] So the purposive behavior of the animal is grounded not in some mysterious *elan vital*, but in its dynamic relationship to its environment. Here animals are seen as constantly absorbed or "captivated" in their environments, driven out beyond themselves, suspended, as it were, between their body and its milieu.

As is clear, Heidegger resisted such a robust account of animals, denying them any kind of interiority or intersubjectivity, unlike in his pre-*Being and Time* work, even though he still speaks the language of hierarchy by placing animals "above" inanimate things. But is this resistance warranted? His phenomenological instincts make him wary of setting forth a full-blown ontology of nature without clarifying fundamental concepts such as life and matter. Yet the momentum of his project for a new ontology both before and in *Being and Time* leads to such questions, and after *FCM*, he never returns to them in adequate detail. What I am suggesting, then, is that his approach should have led him in the direction of a philosophy of nature that engaged the natural sciences in the sense of curbing their ontological overreaches but incorporating their crucial insights. He seems to have been bedeviled by what Ted Toadvine has called "the inherent paradox of any phenomenology of nature":

> [T]to the extent that phenomenology starts from experience, we seem constrained at the outset to reduce nature to the range of our perceptual faculties, to frame it in terms of our spatial and temporal scale, and to encounter it in anthropocentric terms, that is, to humanize it. Nature therefore confronts phenomenology with a problem of transcendence.[42]

Heidegger wants to avoid both positivism and anthromorphism—indeed, in a telling admission in his later Heraclitus course, he says that "human analysis practically runs out of alternatives when it rejects mechanistic views of animality . . . as firmly as it avoids anthropomorphic interpretations."[43] As I will explain in the next chapter, in his later work he appeals to a notion of *physis* that is either a poetic construction or an unknowable thing it itself, a cosmos without a logos, which is to say, not a cosmos at all. In his later writings Heidegger comes at nature more through his notion of "earth," which seems to refer generally to what is traditionally known as the four elements. His idea is that by letting the raw materiality of things shine through, rather than seeing things in terms of human categories, we let beings and nature emerge on their own, and we somehow get more in touch with the "elemental."

One problem with Heidegger's approach in *FCM* is the facile treatment of material things. First of all, he ignores the fact that the stone is not a proper unity; it is a collection or heap of molecules. Obviously the stone has no intrinsic unity, no form or structure; it can be cut in two and remain fundamentally the same thing. But if the molecular structure of the parts composing it changes, then there is indeed a constitutional change going on. Put another way: in his later writings, Heidegger seems to abandon Aristotle's key distinction between artifacts and natural things (specifically, natural things that have structure, form, and intelligibility). While it is indeed obvious that the stone is worldless, it is by no means obvious that the molecules composing it do not possess some particular capacities—perhaps not unlike the "proper way of being" Heidegger attributes to animals—that enable them to enter into certain structural relationships with other beings of their kind and allow them to be taken up and incorporated into organic unities. By forgoing a serious interpretation of the material order, Heidegger blocks the articulation of a continuum view of nature, ontologically dissociating the inorganic and organic realms, just as he severs the human and animal realms.

Interestingly, he wonders about this possibility in an early lecture course: "The question of the extent to which one might conceive the interpretation of *Dasein* as temporality in a universal-ontological way is a question which I am myself not able to decide—one which is still completely unclear to me."[44] As I noted in chapter 3, Heidegger cuts off the temporality of Dasein from any kind of "natural history."[45] The temporality proper to animals or, say, evolutionary history, is deemed beyond the pale of human comprehension, relegated to the realm of the unspeakable. But what if Heidegger extended his novel approach to humans to nonhuman beings? What if, after successfully critiquing the substance metaphysics that had pervaded and perverted the philosophical tradition—as well as its correlative conception of nature—Heidegger had forged a new conceptual scheme for talking about the being of beings which restored not just our subjective experience of nature, but imbued some limited form of interiority to all natural beings, as he did in his pre-*Being and Time* writings on Aristotle? There are glimmerings of such a view. For instance, as Thomas Sheehan explains,[46] in 1928 Heidegger started translating Aristotle's *dynamis* as *Ereignis*, indicating that all natural entities have a proper potentiality of being to actualize or unfold. But this idea is never fully developed. Again, while Heidegger is correct to combat the "tyranny of physics and chemistry" insofar as they reduce the phenomenon of life, this is not to say that they are not genuine regions of being that need to be integrated in a complete account of nature.

III. Critical Considerations

1. Evolution

It is not an accident that the sole monograph devoted to Heidegger's philosophy of life—Krell's *Heidegger and Life Philosophy*—contains not a single entry for either "Darwin" or "evolution" in the index. It is more than a little strange that a philosopher so intent on questioning the foundations of Western thought refused to grapple with what Daniel Dennett has with good reason called a most "dangerous idea," the idea that called into question the scientific, philosophical, and religious foundations of the West: evolution by natural selection.[47] Though evolution in its modern signification was born in England, it found its spiritual home in Germany, where, as Gregory Moore relays, it "achieved the status of a kind of popular philosophy."[48] Indeed, Heidegger's sidestepping of evolution is largely due to his revulsion of the *Lebensphilosophie* and biologism popular at the time; I think it is a safe conjecture that Heidegger considered the enthusiasm for biospirituality a reflection of his own age's *das Man*. From his perspective, embrace of an evolutionary philosophy could only take two forms: a lifeless mechanistic view of nature (the more sober, British iteration, exemplified by Darwin) or a life-drunk vitalism or spiritualism (the more speculative, German iteration, exemplified by Ernst Haeckel). However, dissatisfaction with these interpretations of evolution is not sufficient grounds for ignoring or dismissing the issue altogether.

The question of evolution has received scant attention in recent scholarship on Heidegger's view of life, nature, and animality. As far as I know, Frank Schalow is the only one to make it the main focus of an inquiry. Heidegger's basic position, he writes, is that "while [he] does not embrace Darwinian theory as such, he does not discount our affinity or even ancestry with our animal counterparts. . . . It is not so much the spirit of evolutionary theory which Heidegger opposes, as its pretext of masquerading as . . . materialism."[49] So his phenomenology does not dispute the fact of evolution, only its materialistic interpretation. "Phenomenology," Schalow writes, "serves as a corrective for evolutionary theory by clarifying its presuppositions and preventing the ontic answers it provides about our origins from monopolizing the different approaches."[50] This is fine, as far as it goes, but while phenomenology plays this critical or "negative" role as antithesis to the thesis of positive sciences, two unavoidable questions remain: [I]f the materialistic view of evolution is unfounded, then what is the alternative? And how do humans fit into this scheme? Whatever their limitations, a number of thinkers—including Herbert Spencer, James Mark Baldwin, William James, Darwin himself, and, as we'll see, Nietzsche—attempted to explain how, roughly speaking, consciousness fits within the evolutionary story, without embracing a mechanistic universe. Heidegger does not offer any such positive account; instead, biology and ontology are kept at arms length; and one of his main reasons for doing so is that to cede ontology to biology, or metaphysics to science, would spell nihilism.

Schalow gives three answers for why Heidegger ignored the issue as he did and offers some promising suggestions for how his hermeneutic phenomenology might repurpose evolutionary theory to make it more philosophically palatable. First, Heidegger takes issue with biological discourse because of its "grammar of reference," which overlooks the phenomenon of existing. As Schalow relays, "the designation 'man' as corresponding to a species, is inherently problematic. . . . The individuality of the act of existing itself points to or outlines the meaningful context in which what we can formally indicate to be relevant about our humanity can be addressed."[51] "Species" is thus taken as an abstract concept, a formal indication that, when analyzed, leads us back to facticity. So we should not see evolution, as biologists do, "abstractly in terms of some remote event . . . but rather concretely in terms of its specific relevance to the inquirer addressing these origins."[52] Fair enough, but the danger here is that biology becomes refracted through the human perspective; biology is not first and foremost about "our" existence—it is concerned with the nature of life as such.

The second objection to evolutionary theory—namely, its conception of time—has a similar structure. The timelines according to which evolutionary narratives are constructed are determined by the traditional conception of time as present-at-hand moments. The meaning and ontological significance of evolution are determined by its implications for our future.

Third, evolution is too dependent on the dubious concept of "self-preservation." As Heidegger says, "Darwin's one-sided emphasis on the concept of self-preservation . . . grew out of an economic perspective on man."[53] Though this is arguably a simplistic reading of Darwin, the criticism does show the value of hermeneutics for qualifying evolutionary theory. An *au courant* socioeconomic assumption of the time was smuggled into a biological explanation. For Heidegger, Schalow says, "[animal] behavior can exhibit a modality of concern intrinsic to each organism that cannot simply be reduced to macro forces of the struggle for survival, genetic mutation, and natural selection."[54] All told, Heidegger holds that materialistic and mechanistic evolutionary theory rests on an unfounded body of evidence.

Brett Buchanan also sheds some light on Heidegger's misgivings with evolutionary theory by showing how he was influenced by Uexküll and Karl von Baer, two anti-Darwinists. First, Heidegger thought Darwinism was methodologically naïve in treating the organism in abstraction (in terms of its parts and physiology) rather than as actually encountered (in terms of its movement and behavior). Second, Darwin interprets animals as present-at-hand entities; this causes him to miss their "manner of being" and their "intrinsic relation" to their environment. Baer and Uexküll insisted that organisms have something like an internal structuring principle that guides their development; Baer, for instance, focused on embryology to support this idea. Darwin, however, is held to miss this dimension by focusing on the organism as an object at the mercy of its external environment. As Buchanan points out, Uexküll opposed Darwinism because he saw it "as a 'vertical' model of descent and one that emphasizes far too much a chaotic view of nature's formations. Uexküll was not necessarily anti-evolution, but his focus was . . . toward a more 'horizontal' model that looks at how organisms behave and relate to things across their respective environments."[55] His basic objection was that Darwinian nature had no plan or design and was basically a form of nihilism. According to Buchanan, Heidegger thought Darwinism consisted of "a hopeless confusion . . . both a chaotic freedom (mutation, random accident) and a materialist determinism."[56] In much the same way that Heidegger strove to recover the meaning of being in the age of scientism and technology, Uexküll attempted to recover a view of nature as meaningful in light of Darwinian biology. This purportedly led both to study the phenomenon of animal behavior and neglect the development of species over long stretches of time.

Despite these qualifications, Heidegger's failure to integrate terrestrial evolution into his thought is the greatest gap in his philosophy of nature. In chapter 1, I mentioned the tradition of biophenomenology, represented by thinkers like Jonas and Grene, and throughout I have highlighted parts of Heidegger's corpus with which it resonates, especially the conviction about the need for some sort of return to or retrieval of Aristotle. However, these thinkers were also by and large convinced that the basic insights of phenomenology must somehow be squared with the theory of evolution. Since Aristotle's nature is eternal, his natural kinds (arguably) fixed and unchanging, and his outlook on nature apparently nonevolutionary, a mere revival of Aristotle's philosophy of nature in opposition to modern mechanistic materialism is not sufficient. According to Jonas:

> Aristotle read this hierarchy [the ascending forms of life] in the given record of the organic realm *with no resort to evolution*, and his *De Anima* is the first treatise in philosophical biology. The terms on which his august example may be resumed in our

time will be different from his, but the idea of stratification, of the progressive superposition of levels, with the dependence of each higher on the lower, the retention of all the lower in the higher, will still be found indispensible.[57]

Moreover, the question of a genetic account or narrative of origins is not really avoidable. As I discuss later, Heidegger does offer a nonnaturalistic story about human history—the "history of being"—and it is a wildly implausible one that, as Caputo and Bambach have demonstrated,[58] has decidedly mythological strains. By failing to incorporate the sciences into his overall account, Heidegger left a vacuum to be filled by mythological thinking.

2. Hierarchy

I return now to the question of hierarchy. As we saw above, in his early embrace of a neo-Aristotelian *scala natura*, Heidegger seemed to adhere to something like a "great chain of being"; but this is less clear in *FCM*, where he attacks the language of hierarchy and eventually concedes that the "three level thesis" of stone, animal, and human breaks down. In any case, the issue of hierarchy has received much attention in recent scholarship on Heidegger and animal life. There are those—Krell, Derrida, Agamben—who think that Heidegger's "essentialism" commits him to a notion of hierarchy he claims to eschew and that, rather than pursue an analogical approach to animals, we should yield to their radical alterity, perhaps even abandon the concept of the animal altogether. Others—such as McNeill—think his position is more nuanced in that he rejects any "ontological" hierarchy but allows "ontic" hierarchy. What both camps seem to have in common, however, is the assumption that any kind of hierarchy is anthropocentric and must be avoided because it necessarily depends on some antiquated metaphysical notion of a great chain of being.

I think this assumption is mistaken because it fails to differentiate between healthy and unhealthy hierarchies. Very roughly, in the former, the higher embraces the lower and releases it into a wider communion, freeing it for possibilities it could not actualize on its own; the higher and lower are interdependent, and both orders benefit—life is "enhanced." In the latter, the higher represses the lower and uses it for its own designs; life is exploited or devalued. In short, while there are plenty of unhealthy, socially constructed hierarchies, this does not exhaust the full phenomenon.

Kelly Oliver draws an important connection between Heidegger's views on hierarchy and evolution, insisting that his comparison of humans and animals actually leads to an impasse: "On the ontological level animals and humans have nothing in common and therefore cannot be evaluated in relation to each other. In the end, one main lesson of Heidegger's pedagogical comparison is that any true comparison is impossible."[59] And the original goal of the comparison, recall, was to further illuminate the phenomenon of world. Oliver notes that Heidegger rejects any sort of ladder metaphor to discuss the human/animal relationship. As he writes, "nature . . . is in no way to be regarded as the plank or lowest rung of the ladder which the human being would ascend, thus to assert his strange essence."[60] And yet, as Oliver notes, he goes on to claim that

> the environments of man and animals "are not remotely comparable" because while animals lived under the influence of nature, "man exists . . . out of our own essence and not from nature's influence." With the ladder metaphor he is not so much saying that there is no hierarchy between animals and humans as denying the evolution of the human way of being from the animal way of being. In other words, at least on the ontological level, we aren't related to, or evolved from, animals.[61]

So Heidegger rejects one kind of hierarchy and elects another—and this blocks him, Oliver thinks, from giving any account of our kinship with animals. This kinship cannot be conceived biologically, since this would not do justice to human freedom, transcendence, language, and temporality, yet it cannot be conceived ontologically, since this would project human characteristics onto animals, which, due to our limited access to their way of being, we lack the warrant to do. So when it comes to evolution, Oliver writes, "Heidegger is not so much denying evolution on an ontic level, the level of biologists, as on a conceptual level, the level of philosophers."[62] What does this mean? How are these levels related? This is what leads Krell to speculate that, by positing a sharp essential difference between the human and the animal, "the question of possible stages, steps, or leaps in being will return to haunt [Heidegger]: the ostensibly unified field of *physis* will crack and deracinate in order to expose strata in being."[63]

What Oliver and Heidegger share is the view that any kind of evolutionary approach is necessarily reductive. One of the key issues here is that many of these thinkers think hierarchy ruptures unity of *physis*, whereas I think it secures it; we just have to distinguish between good and bad hierarchies. Heidegger's hierarchy is dissociative because it preserves the distinctness of humans and animals yet fails to account for their kinship; which is to say, he abandons the principle of continuity. But he is right to demand some kind of hierarchy.

3. *The Human/Animal Divide*

Here I will lay out what we can call the "continuity problem" (CP), outline both charitable and critical views of Heidegger's position, and pinpoint the main shortcomings in his approach. This will crystallize the issues at play throughout the chapter and set the stage for Nietzsche's alternative later on.

Bruce Foltz's framing of the CP is useful because he points out that the crucial ontological problem is not mind-body dualism, but mind/life dualism.[64] This is where Heidegger's dissolution of the mind/body problem shifts the ontological spotlight. Foltz thinks the CP comprises two trilemmas: first, the life/nature relation, and second, the human/nonhuman relation. Both problems admit, roughly, three basic solutions, each with its own dangers: 1) physicalism, 2) dualism, and 3) pan-psychism/-vitalism. The danger of the third, Foltz says, is that "nonhuman life will be anthropomorphized, leading us to speak of such things as plant communities and animal language, and thereby tacitly reducing the human to the nonhuman through an incomplete and inadequate understanding of the former."[65] This is almost exactly Heidegger's critique of Nietzsche. But how does Heidegger answer the trilemma? We just saw that he rejects the third option, and obviously he rejects any kind of physicalism. At first he would seem opposed to dualism, given his critiques of Cartesian and substantialist conceptions of mind, yet, as the analysis of this chapter has shown, he retains another kind of dualism with regard to humans and animals. As Tristan Moyle has argued,

> Despite the positive interpretation of animal life [Heidegger takes] from von Uexküll . . . the concept of nature no longer provides the support required for justifying the belief that humans and animals share a specifically natural existence," and "the metaphysical split within human nature simply re-emerges between humans and animals.[66]

Heidegger's claim that animality, life, and the earth are radically other regions of being indicates that these orders are not intelligible, that is, they admit no degree of continuity or analogical relation with the human way of being. Nature cannot be conceived as a "cosmos" in the original sense of an ordered state of affairs. Thus Jonas's original contention—that Heidegger regards humans as aliens adrift in an a-cosmic nature—is sound. Heidegger is a dualist by default.

Some read Heidegger more charitably. McNeill does not see so sharp a dualism in Heidegger's account. He points to a striking quote in which Heidegger speaks of "an intrinsically dominating character of living beings amongst beings in general, an intrinsic elevation of nature over itself, a sublimity that is lived in life itself."[67] He does not seem bothered by Heidegger's ambiguity over the relation between human worlds and animal environments. Though humans are not confined to their own "encircling ring" or perspective, they are nevertheless "transposed in a peculiar way into the encircling nexus of living beings."[68] McNeill softens the human/animal duality by pointing to Heidegger's untraditional understanding of essence: "[T]he claim that the animal is other in essence does not refer to essence in the sense of 'whatness' or substance characteristic of metaphysics, but to the respective ways of Being of human and animal, the kind of presence each displays."[69] But vague recourse to "peculiar" relations and "ways of Being" do not eliminate the problem: how are the two kinds of being related?

Foltz is more specific and thinks that Heidegger does offer a viable conception of life, though he does not specify in which texts or phase of his work this is to be found. He thinks Heidegger's view of life is a retrieval of the Greek notion of life, *zōë*:

> *Zōë* . . . designates a particular character of *physis* within which self-emergence is intensified. *Zōë*, 'life' understood in a Greek manner, designates that which is particularly self-emergent, especially self-unfolding, most prevailing. For the Greeks, *zōë* is understood by means of *physis* as intensified self-emergence.[70]

Contra Krell, Foltz does not think that the problem of life ruptures the unity of *physis* because different orders or ways of being are different "intensities" of emergence and persistence. Moreover, Foltz notes that the Greek view saw life as bound up with soul, *psyche*, and he connects the movement of breathing to the withdrawal and presence-ing of *physis*. Note that Foltz's interpretation is strikingly similar to Heidegger's pre-*Being and Time* account of Aristotle's "levels of soul/life," though he does not cite these texts. While Foltz's account seizes on the most promising aspect of Heidegger's view of living being, I think he does not sufficiently deal with the real ambiguities in *FCM*—indeed, the work is barely discussed in Foltz's monograph on Heidegger's view of nature—and he does not appreciate Heidegger's neglect of psychology and vagary about drives and instincts.[71] Insofar as he elaborated something like Foltz's reading, Heidegger was on the right track, but all told, the charitable readings of Heidegger's view of life are not compelling.

Others are highly critical. Spirited declamations abound. Didier Franck: "The ecstatic determination of man's essence [by Heidegger] implies the total exclusion of his live animality, and never in the history of metaphysics has the Being of man been so profoundly disincarnated."[72] Derrida: "[T]he distinction between the animal and man has nowhere been more radical nor more rigorous than in Heidegger."[73] Agamben: "[Heidegger] is the philosopher of the 20th century who more than any other strove to separate man from the living being."[74] Like Krell, Derrida thinks that Heidegger's view depends on categories it claims to eschew: it aims for "the protection of essences from contamination by lower echelons of beings."[75] It depends, in other words, on traditional categories, most especially spirit. Applying Heidegger's interpretive concept of the "unsaid" to Heidegger's own work, Derrida argues that "spirit" is the unsaid that haunts and subtly sabotages Heidegger's thought and that this is particularly acute in his account of animality. As he writes, "If the world is always a spiritual world . . . if, as Heidegger says at the end of [*FCM*], the three theses, but especially the middle one [that animals are world poor], remain problematical so long as the concept of world has not been clarified, this is indeed because the spiritual character of the world itself remains obscure."[76] The obscurity of spirit is the photographic negative, as it were, of Heidegger's assumption that it even makes sense to seek an "essence" of animality.

While I agree in general with Derrida's rejection of a hardcore metaphysical notion of "animality as such," I do think we need to recognize some sort of stratification among natural beings. There are crude and subtle ways of articulating the spectrum, but there is indeed a spectrum, and our accounts must reflect that. We are right to say that the ways matter, life, mind, and spirit have been ordered and conceived in the past often reflected human biases and blindspots, but we are wrong to conclude that the basic intuition of hierarchy or verticality is false. It will not do to ignore the problem and subvert natural hierarchies by pronouncing different regions as "radically other," as Heidegger does by deeming the third sense of nature ineffable. Again: Heidegger's gesture toward hierarchy and engagement with theoretical biology is a step in the right direction, but his execution is crude.

At the end of *FCM*, Heidegger tells us that he was all along trying to further determine the concept of world. His concessions at the end of the course are revealing:

> An understanding for the fact that there are fundamentally different specific manners of being itself, and accordingly fundamentally different species of beings, was precisely sharpened for us through our interpretation of animality. . . . Animality no longer stands in view with respect to poverty or world as such, but rather, as a realm of beings which are manifest and thus call for a specific fundamental relationship toward them on our part.[77]

In short, the thesis about animals' world poverty is abandoned, as are all the earlier references to animal worlds, intentionality, and even environments. From here on out, world is merely and always spiritual world, and animality will for the most part be passed over in silence as a strange realm subject to the "law of the earth," an ineffable order that, while not quite chaos, cannot be intelligibly spoken of, accessed by, or integrated with the human.

Conclusion: The Kantian Roots of Heidegger's Later Philosophy of Nature: The Earth as the Sublime

Heidegger's apparent eventual position, then, is a refusal to ascribe world and even *Umwelt* to the animal. This is actually most consistent with his thought as a whole. But that locks us into a very limited, and ultimately meaningless, view of nature as *physis*: as a poetic construction for the unity of nature that does not tell us what is being unified, or in what way. I have suggested that though Heidegger was guided early on by a principle of continuity like that found in the traditional notion of the "great chain of being," he apparently abandoned this view in favor of a retrieval of a pre-Aristotelian notion of nature as *physis*. All we can say about it is that animality, life, and the law of the earth are "radically other"; no comparison with the human order is possible.

This is the context in which Heidegger's later, allegedly more nonanthropocentric attitude toward nature plays out: humans are ontologically divorced from animals and all other natural entities. For methodological reasons, Heidegger regards any kind of continuity view as anthropomorphic. Yet this reticence leads him to neglect the crucial category of life that, in his early, pre-*Being and Time* writings, was foundational for his key conceptual innovation, being-in-the-world. As Krell observes, the concept of life haunts Heidegger's corpus, and a biologically informed ontology of life is a road not taken in his later work. We might use Hegel's barb against Schelling as an example: if Heidegger's earlier, neo-Aristotelian ontology of life is like Hegel's picture of clearly defined, progressive stages of Spirit's self-unfolding, then his later, poetic paeans to *physis* are like the night in which all cows are black; in other words, the notion of order, structure, and intelligibility reaching through the entirety of nature, of nature as a cosmos, an ordered state of affairs, is lost.

In concluding this chapter, I want to situate Heidegger within the "hermeneutics of nature," a useful framework Philippe Huneman has recently proposed to understand the evolution of post-Kantian philosophy of nature.[78] This framework can help us see how Heidegger is the heir to the Kantian dualism of "objective" and "sublime" nature. Heidegger's later names for nature (*physis*, the earth, the fourfold) and commitment to the primacy of poetic language for addressing being reflect his view that nature is simply beyond the ken of human concepts and categories, an unknowable thing in itself that can only be addressed through a glass, darkly. This forecloses any attempt to situate human beings within anything like a cosmos, to link them to a deeper animate order, or to incorporate evolutionary theory in any meaningful sense.

By "hermeneutics of nature," Huneman means a way of approaching living and natural being and of conceiving of the relationship between philosophy and science opened up by Kant's *Critique of Judgment*, and that was explored initially by the German idealists and later by phenomenologists. As he explains,

> after Kant, life acquired a specific position within the reflexive examination of our finite thought by itself. Therefore, when Schelling writes that for *Naturphilosophie* "nature is nothing else than the organ of self consciousness," or when Hegel conceives nature as the mind opaque to itself, they both rely on this interpretation of life as a reflexive concept, proper to the necessities of finite thought, elaborated in the Third *Critique*.[79]

In particular, this research project aims to interrogate the "excess" in nature, that dimension of nature that slips the bounds of Newtonian mechanism: "Briefly put, life is in excess of nature pure and simple—epistemologically speaking—and therefore requires another kind of intelligibility."[80] There are three aspects of this project that relate directly to Heidegger's early approach to living and natural being: 1) it does not try to explain life in terms of mathematically determinable laws; 2) it is autonomous from the natural sciences; 3) Huneman notes that, given its commitment to beginning from the meaning/sense of things and its constitution by consciousness, as well as its embrace of 2), phenomenology is especially well suited to the project. The hermeneutics of nature, he says, aims to find "an immanent meaning in nature that is not explicated by the sciences of nature, which are dealing with the laws of nature."[81] Let us take a brief look at the Kantian revolution in the understanding of life in order to expose the Kantian roots of Heidegger's later view of nature.

In the third critique, Kant's investigation of teleology revolves around the phenomenon of the organism. Robert Richards relays Kant's view of organisms:

> [F]or objects to be constituted organisms or as Kant also refers to them, "natural purposes," they have to meet the following criteria: their parts form reciprocal means-ends relationships; those parts come into existence and achieve a particular form for the sake of one another (through growth, maintenance, and reproduction); and the entire system has to be understood as resulting from an idea of the whole. No mere mechanism displays all of these features.[82]

Organisms present a special problem for the Kantian view of nature because they clearly exhibit a kind of order and structure, yet their purposive behavior does not seem explainable by mechanical principles. Kant deems teleological judgments about nature "reflective" rather than "determinate" because they do not involve the application of a universal rule to a particular instance. Determinate judgments, in other words, have a universal and necessary structure that issues from the categories of the understanding. The former class of judgments, Richards writes, are reflective because they

> indicate two related features: 1) that a concept of the whole has to be empirically discovered by an initial examination of the parts; and 2) that such a concept is ultimately grounded not in a necessary requirement of nature—that is, in a natural law ultimately based in the categories—but rather in a necessary requirement of our reflective capacities.[83]

Since such judgments only express regularities, not necessities; they do not reflect the structure of the understanding and cannot in any sense constitute knowledge of the empirical world because they lack the form of universality and necessity. Only mathematical physics possesses this character, which means that, for Kant, biology is not really a science. He declares that since "in each particular natural discipline, one meets only so much real science therein as there is mathematics to be met," there can be "no Newton of the grass blade."[84] That is why he is led to dismiss any attempts at a non-mechanistic biology as nothing but "poetic swooning."[85]

So on the one hand, Kant banishes teleology from natural science. On the other hand, he maintains that we cannot help but understand living things in a teleological manner. But teleological principles cannot explain biological phenomena; we merely must act "as if" they do. They are thus epistemological and regulative, not metaphysical and constitutive. However, Kant accepted the Newtonian view of nature as matter in motion governed by fixed mathematical laws. This is what motivates his dualism of a "kingdom of nature" and a "kingdom of ends." Moreover, as Richards notes, it is because Kant saw the concept of organism as bound up with intentionality that he refused to grant it ontological status: "[T]he principle of the organic itself, because it harbored intentionality, he deemed unfit to be a constitutive principle of nature."[86] So, too, Heidegger rejected a neo-Aristotelian ontology of life and remained agnostic about the intentional life of nonhuman beings.

But what then is to be done about the "excess" of life? For Kant, it migrates to the realm of the sublime, that dimension of nature that exceeds our conceptual and categorial grasp. Richards details this rift in the Kantian view of nature: "While Kant's world of nature clanked on in mechanical fashion, yielding to a jejune understanding, it yet obscured a mysterious, dark *Urwelt*, about which only rumors could be floated."[87] And this *Urwelt* corresponds closely with Heidegger's notion of the earth, the mysterious aspect of *physis*/being that withdraws from human understanding but subtends the world, the horizon of intelligibility. Thus we find recapitulated in Heidegger's thought the same residual dualism between objective nature, which is the province of science, explanation, and theory, and poetic nature, which withdraws from and resists human understanding. Much like Kant, Heidegger holds that natural science is legitimate within certain bounds—it does allow nature to be heard—but that it errs in crossing them and lapses into naturalism, a naïve metaphysics. However, both take as normative a view of natural science that emerged in the seventeenth century; and that view no longer enjoys the scientific consensus it once had.

In sum, Heidegger's early attempts to work out a nonreductive naturalism based on a retrieval of Aristotle's account of life, soul, and nature was eventually derailed by prejudices, inherited from Kant, concerning the natures of (and relationships between) natural science, living being, and human intentionality. This leads us to ask: If Heidegger's philosophy of nature is dubious, is a Heideggerian environmental philosophy, to use his own phrase from another context, a "round square and a misunderstanding"? Before answering this question, we need to take a closer look at Heidegger's later philosophy of nature, which is the basis of most attempts to derive an environmental ethic from his thought.

5

Nature in the Later Heidegger

Earth, Physis, *Technology, Machination, and Poetic Dwelling*

An examination of Heidegger's later thinking about nature is in order, given that most ecological readings of his thought draw on the later texts. In this chapter, I trace the arc of Heidegger's later account of nature through his work in the mid-1930s until the early 50s and focus on 1) his notion of earth, 2) his retrieval of the Greek concept of *physis*, 3) his critique of technology and machination, and 4) his prescription of poetic dwelling in the fourfold. Most of these themes are concerned with the elaboration of the third sense of nature ("primordial" or "poetic" nature) only scarcely sketched in *Being and Time*, and my exegeses are specifically intended to show how they are related to this third sense.

Many approaches to Heideggerian environmental thought see this increasing focus on nature as a sign that Heidegger became a non-anthropocentrist. In translating his critique of metaphysics into a more concrete, historical narrative—the so-called "history of being"—and by casting it in terms of the exploitation of nature, Heidegger *appears* to present himself as a protoenvironmentalist and a nonanthropocentrist. His paeans to poetry, musings on the mystery of the earth, and fascination with *physis* seem to signal a departure from a phenomenological approach to nature in general (the sense nature has for human intentionality) and the allegedly existentialist, anthropocentric slant of *Being and Time* in particular. On this view, Heidegger's turn is a turn toward nonanthropocentrism, or even biocentrism. As I detailed in the first chapter, this view is embraced by the early Zimmerman, Devall and Sessions in *Deep Ecology: Living as if Nature Matterred*, and several others. I think this view is partly mistaken. Though several of Heidegger's later concepts—including his account of "the thing," his critique of cybernetics, the attitude of *Gelassenheit* or "letting things be," and his notion of "authentic use"—are promising for environmental ethics, they are too vague and stray too far from the concrete realities of animal, biological, and natural phenomena. Put simply, despite the adoption of a poetic style and the shift of focus away from humans and toward being, the stubborn conviction about the ontological gulf between humans and nonhumans—as well as the preoccupation with the relation between humans and Being—persists throughout the later work and indicates a residual anthropocentrism. Below, I unpack the main nature-related themes in Heidegger's later work and circle back to the themes traced in previous chapters.

I. Earth

Heidegger's arcane invocations of the earth throughout his middle and later work, though music to the ears of some environmental philosophers, are motivated by several factors, and what we today refer to as the "ecological crisis" is not one of them. Heidegger is not concerned about the depletion of the ozone layer, the extinction of species, global warming, or other environmental problems and policy issues. The destruction of the earth that he laments is a more concrete-sounding locution for the same forgetfulness of being that he bemoans in *Being and Time*. The tragedy has to do with the progressive narrowing of the clearing through which being shines forth, not with increasing impediments to the flourishing of particular beings, the propagation of species, or the health of ecosystems. Even the consequences of the detonation of the atomic bomb itself, we are told, would pale in comparison to a total forgetfulness of being. As such, Heidegger's recollection of the earth should be approached with some skepticism by those looking to enlist him for environmental purposes.

A related point that should give us pause is that Heidegger calls on the earth to fulfill an outstanding problem within his own philosophy. Michel Haar, for one, is convinced that the notion of earth is the basis for the later Heidegger's history of being, according to which history is characterized by a series of "epochs" that are, on the one side, governed by a unique disclosure of being and, on the other, the withdrawal or withholding (*epoche*) of being itself: "Though appearing in the *epoche* and clearing of being as does every being, [earth] is not reduced to a being nor even to the epochal, but it holds itself back, like being, thus preserving an extra-epochal dimension. Historical and yet non-historical, it appears as the most elementary ground of the world."[1] As I discussed in chapter 3, the elaboration of the third sense of nature is left incomplete in *Being and Time*, and the earth must be seen as a means toward filling that lacuna. Whether or not it succeeds in doing so, or whether it is merely a sign that the approach of a fundamental ontology was inherently problematic to begin with, is worth pondering and will be dealt with later on. But the point to keep in mind is that the elaboration of the third sense of nature and the narrative of the history of being occur in tandem. And the latter, as I discuss in the next chapter, is squarely concerned with the problem of nihilism.

Moreover, we cannot ignore the historical and political context in which Heidegger's discussion of the earth takes place. Charles Bambach has documented the serious influence that prominent national socialist intellectuals such as Alfred Baumler had on Heidegger's work in the 1930s, in particular a revival of an archaic Greek *mythos* of the chthonic deities who served as foils to the Olympian pantheon, which represented the forces of light, rationality, and order.[2] Many of Heidegger's contemporaries grafted this struggle between the Titans and the Olympians (or the Dionysian and the Apollonian) onto the contemporary struggle between the Germans and the Enlightenment. John Caputo has, along somewhat different lines, provided a similar exposé on the mythological strains in Heidegger's thinking.[3] Whatever the extent of the influence of these ideas on Heidegger's work, suffice it to say that we should be wary of taking his ideas as pure philosophical reflection on the things themselves, rather than as subject to the cultural, political, and intellectual trends of his own time; ironically, this is entirely in keeping with his own hermeneutical approach to philosophy.[4]

With these caveats in place, let us examine Heidegger's account of the earth. Haar, who offers perhaps the most probing analysis of the notion of earth in Heidegger's work, locates four different senses of the term: 1) as related to and in conflict with world, 2) as an analogue of nature, 3) as the material of the work of art, and 4) as terrestrial home.[5] In this section, I deal mainly with the first two senses of earth. The third is relevant to Heidegger's account of *physis*, which I discuss in the second section, and the fourth will be analyzed in connection with the notion of poetic dwelling and the fourfold, which I discuss in the fourth section.

Heidegger's introduction of the enigmatic notion of earth in the "Origin of the Work of Art" from 1935 initially seems out of place. One would expect the essay to be focused on aesthetics. However, when seen in the context of *Being and Time*, as well as Heidegger's work *On the Essence of Truth* (1931), the simultaneous treatment of the earth and art makes sense. Recall that in *Being and Time* Heidegger neglected to include a description of aesthetic judgment and experience in the existential analytic.[6] In short, the art essay can be seen as Heidegger trying to tie up two "loose ends" from *Being and Time*: aesthetic experience and the third sense of nature as primordial/poetic. As he noted there, the first two senses of nature do not comprise the so-called "romanticist" sense of nature that we readily connect with aesthetic experience. As Haar notes, "It is fundamental that the concept of Earth—absent from *Being and Time* where nature is reduced to a 'subsistent being' . . . is elucidated for the first time in connection with the interpretation of the work of art."[7] Graham Parkes is more specific:

> Any impression that the proper attitude toward things is merely technological is quickly dispelled by this essay, a major concern of which is to describe a way of relating to things that is quite different from taking them as [ready-to-hand or present-at-hand]. The work of art, whose essential nature cannot be appreciated if it is taken as an implement or an object of scientific investigation, is to be seen here as a paradigm of things in general.[8]

Thus it is no surprise that the critique of traditional aesthetics and the inquiry into the ontological significance of art is at once an elaboration of the third sense of nature.

It is much easier to determine what the earth is not than to pin down what it is. First, we can clear away the popular senses. The earth is not the entity that came together some 4 billion years ago. It is not the planet that rotates around the sun and on its own axis, nor is it a planet among other planets. It is not an object or sum of objects. It is not any sort of "prime matter." The earth, in Heideeger's words, "is not to be associated with the idea of a mass of matter deposited somewhere, or with the merely astronomical idea of a planet."[9] Nor is it to be seen as a storehouse of energy for human purposes, a collection of "natural resources." Yet the earth also should not be construed as a gigantic ecosystem or biosphere comprising all organic and inorganic beings, since these are at least partially conceptual frames created by humans to make sense of nature. This is a popular frame for many ecological thinkers. A well-known example is James Lovelock's "Gaia Hypothesis," which construes the earth as a superorganism.

Here, it will be helpful to digress a moment to briefly sketch the history of ecological paradigms. As Michael Zimmerman and Sean Esbjörn-Hargens document, Lovelock's interpretation of nature as a superorganism first arose toward the end of the nineteenth century and underwent a number of metaphorical mutations, from "super-organism," to "economic machine," to "cybernetic web," to "chaos."[10] As they point out, the first person to define ecology, German zoologist Ernst Haeckel, "was inspired by

Darwin's discussion of the 'economy of nature' in the *Origin of Species*."[11] Despite this Darwinian influence, Haeckel hewed to a holistic and vitalistic view of nature, according to which a nonphysical force directed or at least influenced the growth and development of natural beings and was emblematic of a backlash against the materialism implied by Darwin's theory. Moreover, this view of a unified nature already had a long pedigree in German romanticism, stemming from figures like Goethe and Alexander von Humboldt (whose travelogues had a profound effect on Darwin's view of nature). In the early twentieth century, American plant ecologist Frederic Clements employed the superorganism metaphor in order to understand patterns of plant distribution. As Peter Bowler recounts, "Founded within an institutional framework dedicated to practical research on great-plains agriculture, Clements' approach nevertheless represented a direct application of holistic, almost vitalistic, ideas to ecology. The natural vegetation of a region, its 'climax,' had the status of a mature living organism."[12]

After empirical studies cast doubt on Clements' approach, and once the neo-Darwinian synthesis took hold in the 1940s, the superorganism metaphor gave way to the economy metaphor. As Zimmerman and Esbjörn-Hargens note, this metaphor was represented by Henry Allen Gleason, who held that "regions were best described as areas of continual change, competition, and probability, rather than holistic communities."[13] This approach was also embraced by Charles Elton, a student of Julian Huxley (grandson of T. H. Huxley, who is widely known as "Darwin's bulldog" for popularizing the theory of evolution). Elton encapsulates the economic metaphor in the following passage: "The 'balance of nature' does not exist, and perhaps never has existed."[14]

The third metaphor, the cybernetic web, involved the creation of complex mathematical models for mapping and predicting the interactions between organisms and their environment. Eugene Odum, one of its purveyors, "combined the super-organism and economy metaphors into the cybernetic one: an economic-like, self-regulating machine," and "moved the concept of 'ecosystem' into the ecological discourse."[15] He built on the ideas of British ecologist G. Evelyn Hutchinson, who, as Bowler explains, "promoted the view that ecological relationships should be seen as systems governed by causal interactions. . . . The transfer of chemicals and energy through the system is governed by feedback loops that create stability in the face of environmental fluctuations."[16] The basic units of this model are thus not organisms, but quantities of energy, and ecosystems are viewed as more or less efficient distributors of the energy provided by the sun. As Zimmerman and Esbjörn-Hargens note, this metaphor seemed to provide the best of both worlds: holism for romantics and environmentalists, and mathematical precision and predictability for ecologists eager to imbue ecology with the air of scientific authority. For this reason, "ecosystem ecology enjoyed an unrivaled popularity during the 1960s and 1970s and is still the most common understanding of ecology among environmentalists."[17] The fourth metaphor, "chaos," recapitulates the unpredictability of the economy metaphor but with the backing of chaos theory, casting doubt on ecological mainsprings such as equilibrium, balance, and harmony.

Yet as they point out, despite their differences, all of the four ecological metaphors "view 'nature' as a great interlocking order of exterior sensory data," and "all four definitions have been used to exploit the environment."[18] And as Bowler notes, both population ecology and systems ecology view nature in the economic terms of resources and raw material, and both are correlated with the anthropocentric project of improving human management and control of natural processes, the former through free-market mechanisms, the latter through top-down technocratic interventions. While the Gaia framework, often invoked by ecologists and environmentalists bent on combating an anthropocentric view of the earth, would seem to suggest a more holistic, interconnected view of the relationship between humanity and the earth, Heidegger would cry foul, since this still conceives the earth in objective terms as a system whose proper "balance" can be calculated and perhaps even engineered by human ingenuity. Bowler echoes this idea: "[A]s far as systems ecology was concerned, the human economy was simply one aspect of the global network of resource utilization that science hoped to understand and control."[19] As we will see below through Heidegger's notion of "enframing" (*Gestell*), the so-called "holistic view" of nature as an integrated, self-regulating system that is touted as a paradigm shift by many contemporary ecological thinkers is actually the old view of nature as a superorganism, albeit garbed in green drag, an instance of what Heidegger calls "cybernetics" and what we might call "ecologism": the elevation of ecological concepts to ontological status.

For Heidegger, this way of thinking reduces humanity to, at best, an animal species, and at worst, a storehouse of energy, and thus glosses over its defining characteristic, namely, its openness to being. Forgetfulness of being is thus very much on par with ignorance of the earth. Unsurprisingly, the earth escapes the intentional stances of the ready-to-hand and the present-at-hand. So long as we one-sidedly conceive of beings either as equipment or as objects—objects in themselves or in a system, atomism or holism—we pass over both their (and our) essential connection to the earth and the earth itself, much like the phenomenon of world is passed over in average everydayness and the theoretical attitude as discussed in *Being and Time*.

So much for what the earth is not. To get at what it is, we need to go through Heidegger's interpretation of the artwork, and this for three reasons. First, the encounter with the artwork will for Heidegger become the paradigm for the genuine encounter with beings.[20] Second, the genuine encounter with beings will become more and more closely associated with the third sense of nature. Third, Heidegger's analysis of technology concludes with the suggestion that the way for humanity to twist free of the grip of *Gestell*—and thus save and properly dwell on the earth—has much to do with a renewed relationship to art.[21]

To begin, Heidegger is emphatic that, as it were, the essence of the artwork is nothing aesthetic: "Reflection on what art may be is completely and decidedly determined only in regard to the question of Being. Art is considered neither an area of cultural achievement nor an appearance of spirit."[22] Since his interest is in the being of the artwork, he approaches it first of all as an entity, and since his approach is phenomenological, he immediately brackets the common conceptual frames in which artworks and "aesthetic objects" are interpreted. Hence Heidegger begins the essay offering three common interpretations of "the thing"—as substance, as sense-manifold (*aistheton*), and as form-matter composite—only to conclude that they overlook what is essential.

The first interpretation of the thing frames it as a substance. Heidegger observes that this "definition of the thingness of the thing as the substance with its accidents seems to correspond to our natural outlook on things" and says, "Obviously a thing is not merely an aggregate of traits, nor an accumulation of properties by which that aggregate arises. A thing, as everyone thinks he knows, is that around which the properties have assembled."[23] Heidegger then suggests that this tried and true formula does not grant the thing its "independent and self-contained character." It could be that, in order to orient ourselves and feel secure in an unfamiliar world, we impose on things a neat scheme of classification into substances and accidents that we then take for granted as the way things are.

The second interpretation holds that a thing is "that which is perceptible by sensations" and is "nothing but the unity of a manifold of what is given in the senses."[24] But here Heidegger questions the concept of "sensation" or "bare sense impression" and makes the phenomenological point that "[w]e never really first perceive a throng of sensations, e.g., tones and noises, in the appearance of things . . . rather we hear the storm whistling in the chimney, we hear the three-motored plane. . . . In order to hear a bare sound we have to listen away from things, divert our ear from them, i.e., listen abstractly."[25] In trying to cease foisting our own concepts on the thing, we inadvertently cloak it in other concepts, this time under the rubric of "sensations" and "bare sense data." Empiricism, in other words, turns out to be another, more insidious abstraction. So Heidegger dismisses both interpretations: "Whereas the first interpretation keeps the thing at arm's length from us, as it were, and sets it too far off, the second makes it press too physically upon us. In both interpretations the thing vanishes."[26]

According to the third interpretation, Heidegger says, "The thing is formed matter. This interpretation appeals to the immediate view with which the thing solicits us by its outward appearance (*eidos*). In this synthesis of matter and form a thing-concept has finally been found which applies equally to things of nature and to utensils."[27] Here artifactual being is transposed to natural being *tout court*; everything natural is taken to be produced for a purpose, to be made for a use: "Usefulness is the basic feature from which this being regards us. . . . A being that falls under usefulness is always the product of a process of making. It is made as a piece of equipment for something."[28] This interpretation of the thing underwrites the so-called teleological view of nature, which we tend to assume was supplanted by a mechanistic view in the modern period. Yet the key to Heidegger's analysis here is his conviction that something survives the conceptual changing of the guard. That something is what Zimmerman has termed "productionist metaphysics," the assumption that to be is to be produced for a purpose and that humans are essentially producers. The objective sense of nature thus comes to be seen as another, more aggressive form of the productive sense, because it is ultimately taken with a view to application. Modern technology was not made possible by modern science, but precisely the reverse. In making this move, Heidegger is driven to seize upon a narrative thread running throughout the whole of Western metaphysics, from the Greeks up through Nietzsche, and the thread he chooses is precisely this instrumentalist ontology or productionist metaphysics. What Heidegger is saying is that in *Being and Time* he, too, was a partial prisoner of the prejudice of productionist metaphysics. The existential analytic is compromised because its starting point is prescribed by the regnant interpretation of being of the modern age, namely, technology or what he will call machination. The motive for investigating the relationships between art and production is to recapture a more fundamental sense of nature.

Heidegger's strategy is to displace the traditional framework of aesthetics and unmask it as a symptom of the forgetfulness of being discussed in *Being and Time* and the blindness to the essence of truth as discussed briefly in the latter and extensively in the 1931 lecture, *On the Essence of Truth*. He is convinced that the poverty of modern aesthetics lies in its dissociation from truth, while the poverty of modern theory of knowledge is that truth has been dissociated from being. Thus the experience of the work of art is to be seen as a paradigm for experience in general. He claims that art should be encountered as the "setting to work" of truth or the "happening of truth," a notion later referred to as *Ereignis*. Despite these many unwieldy locutions, his basic point is that the fundamental sense of truth is ontological, not epistemological; it is the sheer arising, appearing, manifesting of being, not the correspondence of a mental representation and its object. Julian Young offers a helpful schema that clarifies the relations between earth, world, and truth: "[Heidegger's account of truth] is a complex of four elements; the undisclosed (Earth), the disclosed (world in the ontic sense), the horizon of disclosure (world in the ontological sense), and man, the discloser."[29]

But how does the artwork figure in this constellation? Since truth is now taken as a process of revealing and concealing (rather than a property of judgments or propositions, or the correct mental representations of a subject), and since the truth process occurs for us most basically as being-in-a-world, art must be approached in its connection with world. And whereas the two modes of comportment carefully catalogued in *Being and Time*, the ready-to-hand and the present-at-hand, were correlated with senses of nature (the productive and the objective, respectively) that were wholly "intraworldly," this new comportment, the encountering of the work of art in its being and truth, promises both to expand the notion of world and to supplement it with a correlative sense of nature, one, moreover, that is not intraworldly, that escapes the ambit of intentionality and intelligibility. The earth refers, in Young's phrase, to "the dark penumbra of unintelligibility that surrounds . . . our human existence."[30]

The earth is broached mainly through two examples in the text: van Gogh's painting of a pair of shoes and a Greek temple. In the first example, Heidegger offers an interpretation of the world of the apparent owner of the shoes, a peasant woman. But the point of his analysis is to show that the object (the shoes) implies a vast network of meaningful relations, what in *Being and Time* he called a "totality of significance," in short, a world. The world does not appear as an object in the painting, but we can only understand and gain access to the painting at all because there is some degree of overlap between the scene depicted there and our own world. At first, this seems of a piece with the analysis of equipment in *Being and*

Time, yet Heidegger includes something new: the earth. He claims that not just world, but earth is implicated in and thus partly constitutive of the "manifest content" of the painting:

> The equipmental being of the equipment consists in its usefulness. But this usefulness itself rests in the abundance of an essential Being of the equipment. We call it reliability [i.e., readiness-to-hand]. By virtue of this reliability the peasant woman is made privy to the silent call of the earth; by virtue of the reliability of the equipment she is sure of her world. World and earth exist for her, and for those who are with her in her mode of being, and only thus—in the equipment. . . . The reliability of the equipment first gives to the simple world its security and assures to the earth the freedom of its steady thrust.[31]

Despite the notorious obscurity of this passage, a few things are clear. First, the mention of "reliability" and its fragility reminds us of the discussion of equipmental breakdown in *Being and Time*. Natural materials resist our attempts to fashion them to our purposes, and our purposes only take form as over against such resistance. So the notion of earth is very much prefigured in the discussion of equipment in *Being and Time*, where Heidegger remarks that equipment refers to the natural material from whence it came.[32]

Second, world is associated with security, order, and stability, while earth connotes freedom, violence, and disruption. The latter call up the discussions of anxiety and freedom toward death in *Being and Time*, the total breakdown in the network of meaningful relations and the subsequent confrontation with the nothingness of both self and world. The following two quotations from both texts sharpen the similarity: "Earth shatters every attempt to penetrate it. It causes every merely calculating importunity upon it to turn into a destruction. This mastery and progress of technical-scientific objectification of nature . . . remains an impotence of will."[33] Yet in *Being and Time*, it is the world, not the earth, that fills the role of the void, the nothing, the nullity on which the self shipwrecks: "[T]he world as such is that in the face of which one has anxiety."[34] In the earlier works, the breakdown of meaning is located much more on the self's experience of the nullity of the world; it is thus presented as a more subjective and worldly event. In the art essay and afterward, however, it is construed less subjectively and as more extraworldly. As Jacques Taminiaux notes, in *Being and Time*, *physis* "is not the earth in conflicting relationship with the world, because the world according to fundamental ontology is not built upon *physis*. At this point in Heidegger's itinerary, *physis* is not at all an enigmatic source regulated by the tension of unconcealment and reserve. Instead, as soon as it appears, nature is . . . intraworldly."[35] The analysis of equipment leads us to the notions of world and earth, and these can only become available to us, Heidegger thinks, through the medium of the work. This is why the artwork is granted such importance: "To be a work means to set up a world. . . . The work holds open the open region of the world."[36] Here he invokes the example of the Greek temple:

> It is the temple-work that first fits together and at the same time gathers around itself the unity of those paths and relations in which birth and death, disaster and blessing, victory and disgrace, endurance and decline acquire the shape of destiny for human being. The all-governing expanse of this open relational context is the world of this historical people.[37]

The work raises a world in the sense that it erects and opens up and serves as the reference point for a horizon of meaning in which things can show up in their distinctness and in connection with each other. This horizon thus corresponds to the aspect of unconcealment, manifesting, and presencing proper to truth, an aspect Heidegger around this time begins to frequently refer to as *physis*. But this opening only occurs against the limiting background of earth: "The Greeks early called this emerging and rising in itself and in all things *physis*. It illuminates also that on which and in which man bases his dwelling. . . . Earth is that whence the arising brings back and shelters everything that arises as such. In the things that arise, earth occurs essentially as the sheltering agent."[38] Heidegger is explicit about the connection between the artwork and the intimate relation between world and earth: "The setting up of a world and the setting forth of earth are two essential features in the work-being of the work."[39] This is why Heidegger wants to frame the artwork not purely in terms of aesthetics, but as "truth setting itself to work": because there are revealing and concealing functions to it that are in perpetual tension with each other. As he says, "The world, in resting upon the earth, strives to surmount it. As self-opening it cannot endure anything closed. The earth, however, as sheltering and concealing, tends always to withdraw the world into itself and keep it there."[40] The work of art is, as it were, the portal through which the creative conflict of world and earth is manifested and the cradle in which it is preserved, in which the opaque and undifferentiated forms of the earth separate and congeal and form a world. But again, no mystical, exclusive importance need be attributed to artworks here; all that is meant is that, because they stand outside the contexts of utility and objectivity, aesthetic phenomena are more likely to "tip us off" and "clue us in" to what is always already going on with any phenomenon: namely, its sheer coming to presence and withdrawal.

By stressing the intimate connection between world and earth, revealing and concealing, Heidegger is trying to counteract the assumption that the earth is purely unintelligible and chaotic, bereft of form, order, law, and limit. Hence he says that earth is "sheltered in its own law. . . . Earth, bearing and jutting, endeavors to keep itself closed and to entrust everything to its law."[41] Creation is not the one-sided imposition of form on mere matter, but the development of latent potentials already nascent and lurking in the Earth. Heidegger appears to have developed this idea in his 1931 lectures on Aristotle's metaphysics, in which he analyzes Aristotle's concept of *dynamis* (potency or power).[42] Haar comments: "[The earth] is the reserve of possible forms to which manifestation would only give body. . . . Earth is not chaos . . . Earth is for Heidegger a secret sketch of forms."[43] Thus art, for Heidegger, "is the disclosure in works of forms not yet sketched but secretly prefigured." He sums it up thus: "There lies hidden in nature a rift-design, a measure and a boundary and, tied to it, a capacity for bringing forth—that is, art."[44] So art is taken neither as the imitation of nature nor as the superimposition of human forms and modes of perception onto it, but as furthering nature's own possibilities. Obviously, this points in the direction of a less exploitative relationship to beings. By remaining open to their own peculiar possibilities and letting them come forth just as they are, we can assist rather than stymie their unfolding and flourishing.[45] Let me briefly note here that this idea harks back to the pre-*Being and Time* works on Aristotle and seems to be a promising idea for an environmental ethic.

While the earth tries to fill in a gap left by the treatment of nature in *Being and Time*, it only refers to the realm of concealment, an extraworldy, semihistorical order that, though it cannot be simplistically regarded as chaos or abstractly taken as a reservoir of "prime matter," it does not canvass the realm of unconcealment that is worldly, yet is populated by animals, plants, and other life forms. As soon as these beings fall within the ambit of language, world, and meaning (the realm of unconcealment) their essential, primordial, natural being is covered up. "But," Haar observes, "nowhere does Heidegger consider that the very being of natural beings is exclusively derived from the world. From this point of view he is a 'realist.' "[46] It is just that we can never clarify the actual nature of these beings, and natural sciences such as biology and ecology fool themselves into thinking they can do so. Heidegger seems to regard all nonhuman beings as belonging to the earth. In the art essay, he says that "a stone is worldless. Plant and animal likewise have no world; but they belong to the covert throng of a surrounding into which they are linked."[47] Is this latter phrase, whose meaning is hardly clear, synonymous with the earth? It seems so. Earth comprises the being of animals, plants, and so on, the being of life, but also exceeds it, since it refers to the inorganic realm as well. It is a kind of limbo between the human world and the self-concealing dimension of being. Heidegger is forced into this position because he rejects an approach to natural entities based on sympathy or overlapping characteristics.[48]

It is worth noting here that Heidegger must maintain a clear ontological line between world and earth because of his resistance to think of humans as animals, or more specifically, as organic bodies. As he says in the *Letter on Humanism*, "the human being is essentially different from an animal organism."[49] However, he is not here denying that humans are embodied *in some way*. Hence the important distinction, as Caputo points out, is not between human being and animal organism, but between lived body and organic body.[50] The point here is that there may be room to push world down beneath the human plane and concede some sort of capacity for world-disclosure, language, and "culture"—some degree of interiority— for nonhuman beings. As we saw in previous chapters, Heidegger flirted with such a nonreductive naturalism earlier in his oeuvre, but his reticence about dragging humans down into the welter of animality and disregarding being stays his hand from situating humans along a common ontological continuum with nonhuman beings.

II. *Physis*

Heidegger's first sustained treatment of the Greek term *physis*, found in his 1935 *Introduction to Metaphysics*, is accompanied by two other new and pivotal themes that come to dominate his work into at least the early 1940s: a confrontation with Nietzsche's thought and the history of Being. While Heidegger elsewhere seeks to ground his interpretation of *physis* in an actual ancient text (Aristotle's *Physics*), it is vital to see that here the retrieval of *physis* is advanced as the foundation of a critique of the entire history of Western metaphysics, that this history is said to culminate in the philosophy of Nietzsche, and that the overcoming of this history depends in no small part on the articulation of a new account of nature. In this section, I trace Heidegger's retrieval of *physis* through *Introduction to Metaphysics* and "On the Essence and Concept of *Physis* in Aristotle's *Physics B*" in order to further flesh out his attempt to furnish a third sense of nature.

1. Introduction to Metaphysics

Heidegger begins his discussion of *physis* by offering a definition: "[*Physis*] denotes self-blossoming emergence, opening up, unfolding, that which manifests itself in such unfolding and perseveres and endures in it; in short, the realm of things that emerge and linger on." He continues: "*Physis* is the process of arising, or emerging from the hidden, whereby the hidden is first made to stand." Finally, he flatly asserts, "*Physis* is being itself."[51] The latter is, needless to say, a bold statement. But Heidegger only compounds the confusion, announcing in another text that "Truth belongs, as self-disclosure, to being itself: *physis* is *aletheia*, disclosure."[52] So *physis*, being, and truth all refer, more or less, to the same thing. This is a dizzying constellation of decisive terms, but before unpacking it we need to lay out more of its context.

Second, Heidegger suggests a semantic kinship between *physis* and "phenomenon" by comparing the roots of the two words: "Recently the root *phy-* has been connected with *pha-phainesthai*. *Physis* would then be that which emerges into the light, *phyein* would mean to shine, to give light and therefore appear."[53] Moreover, he says, "The radicals *phy* and *pha* name the same thing. *Phyein*, self-sufficient emergence, is *phainesthai*, to flare up, to show itself, to appear."[54]

Third, Heidegger explains how an echo of this word still sounds in our modern era in the form of "physics" and the philosophy of materialism and notes how the original meaning of the word is covered up by the latter view:

> *Physis* originally encompassed heaven as well as earth, the stone as well as the plant, the animal as well as the man, and it encompassed human history as a work of men and the gods. . . . [If] the motion of material things, of the atoms and electrons, of what modern physics investigates as the *physis*, is taken to be the fundamental manifestation of nature, then the first philosophy of the Greeks becomes a nature philosophy, in which case all things are held to be of a material nature. . . . But this narrowing of *physis* in the direction of physics did not occur in the way that we imagine today. We oppose the psychic, the animated, the living, to the physical. But for the Greeks all this belonged to *physis* and continued to do so even after Aristotle.[55]

As I noted above in connection with the earth, the dominant and decisive interpretation of a thing as a "product" set the stage for the Western metaphysical tradition. In Christian philosophy, this interpretation was given a different spin: to be a thing was to be created by God. In modern philosophy, through Descartes, the thing is taken as a positing of the subject, as a lifeless, extended stuff. What Heidegger is pointing to is not merely the ontological severance of body and *mind*, but the dissociation of body and *life*, the draining of *anima*, *psyche*, and *bios* from *physis*.

Heidegger then makes the provocative claim that the misinterpretation of *physis* is the hidden logic of Western metaphysics: "[P]hysics has determined the essence and history of metaphysics. Even in the doctrines of being as pure act (Thomas Aquinas), as absolute concept (Hegel), as eternal recurrence of the identical will to power (Nietzsche), metaphysics has remained unalterably physics."[56] Heidegger reads in—or reads into—the great thinkers of the West a failure to adequately address the being of nature, which is by this point the same as the nature of being. Hence Nietzsche's inversion of Platonism and attempts at a renaturalization of man and revaluation of all values carry special import for Heidegger, since they question the conceptual foundations of Western thought. The *Introduction to Metaphysics* thus points forward to the Nietzsche lectures as the culmination of Heidegger's determination to drill down to the foundations of Western thought and his growing recognition that doing so demands a new account of nature, or perhaps the recovery of an old account of nature.[57] While Heidegger was never explicitly determined to work out a "philosophy of nature," his pursuit of the question of being led him down that path.

It is thus no surprise that Heidegger links *physis* to the emergence of world and the reduction of *physis* described above to the eclipse of world: "[T]his power [i.e., *physis*] first issues from concealment, i.e., in Greek: *aletheia* (unconcealment) when the power accomplishes itself as world."[58] He links this to the Heraclitean notion of *polemos*, of beings as the children of strife.[59] This recalls the discussion in the essay on the work of art of the artwork as the site of strife between world and earth. The reduction of *physis* is also bound up with the collapse of the world: both involve the disclosure of beings as objects, which in *Being and Time* is described as the breakdown of the totality of significance that "de-worlds" objects. Heidegger lists the "destruction of the earth" as one of the main elements of world darkening. As I detail below, the project of modern technology is bent on world domination, to erect a completely self-sufficient system impervious to and independent of the power of the earth, yet it fails to realize that the earth is the secret source of its power and the basis of its being.[60] The description of the earth in the art essay as the repository of all possible forms, a "secret sketch of forms," is replaced by a picture of nature bereft of form, *telos*, and meaning, as a platform for human projects, a reservoir of resources, and a supply of energy.

Much of this is important for setting up Heidegger's account of technology. But beneath the creative etymology and exegetical revisionism we should descry a message of cultural and spiritual revival through the recovery of a pristine relation to nature. That is, the root of nihilism is here taken to be the lack of an original relation to nature. Whatever Heidegger's real or feigned divergences from national socialist ideology, it simply cannot be denied that he at least subscribed to the categorically antimodern view of historical decline, as well as its flipside, the destined revival. The *Introduction to Metaphysics* is very much a product of its time, despite Heidegger's avowals therein that philosophy is necessarily untimely; the text is redolent with its contemporary cultural and political atmosphere and explicitly references national socialism.[61] While this is not grounds for dismissing the philosophical import of Heidegger's introduction of *physis*, we should look at a more specific, textually supported discussion of *physis* unencumbered by such sweeping, apocalyptic rhetoric and "crisis consciousness." Heidegger's 1939 lecture on Aristotle's *Physics B* is an ideal candidate.

2. "On the Essence and Concept of Physis *in Aristotle's* Physics B"

Heidegger is hardly subtle about the importance he attributes both to *physis* and to Aristotle's most extensive treatment of it. Near the outset of his own investigation, he declares that "Aristotle's *Physics* is the hidden, and therefore never adequately studied, foundational book of Western philosophy."[62] Heidegger aims to tease apart the concepts that become pivotal for the tradition and continue to exert a hidden pressure on the present: form, matter, motion, work, actuality, *dynamis*, energy, and entelechy. He is convinced that *physis* is, as it were, the soil out of which these crucial concepts first emerged, later congealed, and eventually calcified, their sense losing touch with their origin.

By this point, he has also already come to key conclusions about the right and the wrong way to think about nature. Holderlin has become the warrior poet of Greco-Germanic purity, the guardian of little things.[63] Heidegger inserts an excerpt from one of his poems right at the start of the Aristotle essay, praising what he takes to be Holderlin's equation of nature and being. This is significant because it indicates that Heidegger has concluded that the proper way of encountering nature is through a poetic way of thinking, speaking, and seeing that he will come to call "mindfulness," "meditative thought," and "poetic thinking" (as opposed to "calculative thought") characterized by the disposition of "releasement" or *Gelassenheit* (as opposed to the "will to will"). Nietzsche, on the other hand, has by now been cast as the final avatar of metaphysics and the herald of the atomic age. According to Nietzsche's vision of a renaturalized humanity,

> *homo naturae* is someone who makes the "body" the key to the interpretation of the sensible world and who thus secures a new and harmonious relation to the "sensible" in general . . . to the passions and drives and whatever is conditioned by them. . . . In virtue of this new relation these people bring "the elemental" into their power and by this power make themselves capable of the mastery of the world in the sense of a systematic world-domination.[64]

Heidegger's Nietzsche calls not for an over-man, but an under-man, a regression to an animalistic state of being beholden only to its own instincts, oblivious to the call of Being, completely closed in on itself and drawing all else within the myopic ambit of its own drives. It is important to keep in mind that at the time of this treatise on Aristotle, Heidegger's Nietzsche lectures are nearly complete and, reading Nietzsche through Ernst Jünger, he is convinced that Nietzsche's vision is not the antidote to nihilism, but its poison perfectly distilled.[65]

Holderlin, however, whose account of nature Heidegger lectured on in 1934, is on the right track. For Holderlin, "'Nature' becomes a word for 'being.'"[66] This is Heidegger's point of departure for the interpretation to follow, and underwrites his claim that Aristotle's *Physics* is the book that sets the pace for the tradition.

There are three main themes or tactical moves Heidegger makes in the Aristotle interpretation that we need to look at: motion, presence, and form and matter. For Aristotle, the defining characteristics of beings that are "by nature" are, first, being in a state of change or motion, and second, possessing the principle and end of that motion in themselves. A natural being brings forth (*physis*) its change/motion of its own accord, while the change/motion of an artifactual being is brought forth (*techne*) by an external agent. This begs the question of the meaning of motion and movement. Heidegger is first of all concerned to bracket any prejudices we might have about this concept: "We of today must do two things: first, free ourselves form the notion that movement is primarily change of place;

and second, learn to see how for the Greeks movement as a mode of Being has the character of emerging into presence."[67] The conviction is that the shades of meaning attaching to our modern understanding of movement overshadow something ontologically definitive, and that this is a symptom of our impoverished understanding of nature. So one of Heidegger's first moves is to draw a distinction between "movement" and "movedness." While movement is the opposite of rest, movedness comprises both movement and rest. Movement has an ontic significance: it refers to an accidental feature of beings that can come and go, but it does not make them be what they are. Movedness has an ontological significance; it is constitutive of the beings to which it is ascribed.

Heidegger also examines movement in Aristotle's rejection or *Aufhebung* of the doctrine of Antiphon, whom Heidegger seems to be framing here as the founding father of materialism and the modern misconception of motion. This is where he makes the connection between motion and presence. There are four main parts (and problems) in Antiphon's approach. First, the ontological sense of motion is lost on him: "[A]ccording to [Antiphon's] understanding of *physis*, all character of movement, all alteration and changing circumstantially devolves into something only incidentally attaching to beings. Movement is unstable and therefore a non-being."[68] Second, as a corollary of the former claim, being is conceived of as that which is stable, as substrate, as substance that suffers the slings and arrows of movement. Third, being, the stable, is interpreted as constancy, as "present-ness." Here, Heidegger makes a distinction that parallels that between movement and movedness: present-ness and presencing. The former is static, the latter dynamic. Heidegger thinks these two ways of regarding beings are reflected in our language, as parts of speech. We conventionally refer to things as nouns, which sometimes act or are acted upon, but normally are simply there, given, present. Yet Heidegger's phenomenology is intent on unsettling this grammatical and ontological prejudice and letting things "oscillate," come forth, to encounter their emergence in their enduring, their origin in their occurring, their what-ness in their that-ness; this is what he means by presencing. The fourth and final part of Antiphon's doctrine is that it "understands the being of [*physis*] via a reference to 'beings' ('the elemental)."[69] In other words, it is ontologically shallow; it takes a particular kind of being—"the unformed that is primally present," that is, that which underlies, the substrate—and makes it definitive for all beings. It is not hard to see why Heidegger thinks that this same kind of thinking fuels modern materialism. If particular beings are all that exist, then those that most truly exist are those that underlie and compose everything, those whose existence can be quantified, objectified, and verified by experimentation. Thus, all that "really" exists is atoms, subatomic particles, quarks, and so on.

The common root of these problems is that they disregard what in Aristotle is traditionally referred to as the doctrine of substantial change, which concerns the coming to be and passing away of substances. Hence, Heidegger quotes Aristotle's reference to two ways of speaking about *physis* as "self-placing into appearance" and "privation." His strategy is to show how form comprises the two themes mentioned above—movedness and presencing—yielding a much richer concept than the traditional view of form as the source of stability and structure in a natural being. To do this, he must show how form is related to idea and *eidos*. His first move is to chastise Plato:

> Idea is "the seen," but not in the sense that it becomes such only through our seeing. Rather, idea is what something visible offers to our seeing; it is what offers a view; it is the sightable. But Plato, overwhelmed as it were by the essence of *eidos*, understood it as something independently present and therefore as something common to the individual "beings" that "stand in such appearance." In this way individuals, as subordinate to the idea as that which properly is, were displaced to the role of non-beings.[70]

Aristotle, however, cleaves closer to our actual experience of beings, giving more ontological weight to the perceptible individual and holding that views only come to us in and through appearances. From here, Heidegger translates form as "placing into the appearance." He contends that this broader, more experientially grounded notion of form encompasses the themes of movedness and presencing discussed earlier.

As for matter, Heidegger's reinterpretation here parallels what he said regarding the earth in the art essay. Matter is not to be seen as an indeterminate, formless stuff, *a la* the notion of prime matter. In the standard interpretation, matter comes to mean "material for production," the blank slate on which forms are imposed. Yet, Heidegger asks, "What does 'matter' mean? Does it mean just 'raw material'? No, [according to Aristotle, it] means the capacity [*dynamis*], or better, the appropriateness for. . . . The wood present in the workshop is in a state of appropriateness for a 'table.' "[71] It should be seen as in a state of potentiality, as harboring possible forms within it. The earth is not primarily "there for the taking," but rather "there for the giving": "that 'whence' self-emerging, self-unfolding, and self-opening arise and 'unto which' they recede."[72]

Heidegger ties all of this together by unmasking a hidden side to the major terms in his interpretation, a side neglected by the tradition and even by Aristotle himself. Movedness is related to *telos*, presencing to absencing, and form is taken as twofold, carrying within it privation as a positive phenomenon. Heidegger invites us to stop thinking of *telos* as goal or purpose: "*[T]elos* does not mean 'goal' or 'purpose,' but 'end' in the sense of the finite perfectedness that determines the essence of something; only for this reason can it be taken as a goal and posited as a purpose."[73] The end should not be taken as something "outside" of a thing, toward which the thing is moving, just as movement should not be taken as an accidental aspect of an object moving through space toward a fixed point. Rather, the end should be grasped as already contained within the thing and as constraining its movedness and unfolding: "[T]he movedness of a movement consists above all in the fact that the movement of a moving being gathers itself into its end, *telos*, and as so gathered within its end, 'has' itself."[74] This idea is captured by the term *entelechy*. He then connects this to the notions of *energeia* (actuality) and work: "In Greek thought energy means 'standing in the work,' where 'work' means that which stands full in its 'end,'" by which he means the work of nature, not art.[75] Here, Heidegger is contrasting two different modes of encountering things we discussed above: productively or poetically, instrumentally or aesthetically. When we take a thing as being for a certain use or purpose, we level off its own unique mode of presencing and draw it within the fixed ambit of our own projects; we definitize and deform it. Thus is the advent of the modern interpretation of movement, which construes nature, in Whitehead's memorable phrase, as "a dull affair, soundless, scentless, colourless; merely the hurrying of material, endlessly, meaninglessly."[76]

Next, Heidegger picks at the view that privation is a mere negation, an absence of form or lack of being, suggesting that it be regarded as a positive phenomenon. A thing in a state of privation should be seen not as a present-at-hand actuality simply lacking or deficient in something proper to it, but as caught in a tension between *dynamis* and *energeia*, a coming futural state and an already actualized present state. He quotes Aristotle's admission that "privation too is something like appearance" and argues, "When something is missing, the missing thing is gone, to be sure, but the goneness itself, the lack itself, is what irritates and upsets us, and the 'lack' can do this only if the lack itself is 'there,' i.e., only if the lack is, i.e., constitutes a manner of being."[77] "Absencing," he continues, "is not simply absence; rather, it is a presencing, namely that kind in which the absencing (but not the absent thing) is present."[78] Absencing is here taken not as the failure of consciousness to completely determine its object, but as a manner of being of thing's themselves, part of the grain of nature. In a way, the interpretation is bent on arriving, through a rigorous investigation of Aristotle's treatise, at Heraclitus's statement that *physis* loves to hide.

Finally, this constellation of concepts is presented as the basis for a new, bivalent understanding of *physis*: "The act of self-unfolding emergence is inherently a going-back-into-itself."[79] This means that what is given, what comes forth, is inherently incomplete, not fully present, and not fully actualized. This aspect of *physis* appears to correspond to the earth, although, as noted above, earth is not simply concealment, while *physis*, as we just saw, is not merely unconcealment. Heidegger even claims, at the end of the Aristotle treatise, that "*physis* is *aletheia*," that is, the process of concealment and unconcealment.

With the conceptual and historical foundations of Heidegger's interpretation of *physis* in place, we can now look at his diagnosis of the modern age.

III. Technological Nihilism

The originality of Heidegger's philosophy of technology is due in part to his claim that it is not modern science (specifically, the mathematical projection of nature in modern physics) that makes modern technology possible, but precisely the reverse. From the standpoint of common sense, technology is simply "applied science." However, when he sets his sights on the phenomenon of modern technology, Heidegger treats it in the same way he studied everydayness in *Being and Time*: by bracketing common theoretical and philosophical assumptions about the phenomenon and tracing the historical development and experiential origin of its conceptual constituents. As such, Heidegger traces the roots of modern technology back to Greek life and metaphysics. Thus his claim is that technology underpins not just science, but also modern, medieval, and ancient metaphysics.[80] It is commonly assumed that the theoretical framework of natural science must be put in place before scientific knowledge can be acquired and then applied in technology; theory precedes practice. But by defining technology in terms of its intentional structure, rather than in instrumental or practical terms, Heidegger is saying that for the standpoint of natural science to emerge, there must first be a technological standpoint, and he will characterize this as a will to master nature. The reinterpretation of reason in modernity is actually a project of the will, a project he thinks was already underway in

Greek metaphysics. The ancient assumption that being is presence is now understood as the technological will to master nature, including human nature. The way forward is the "step back" from metaphysics, a move that involves both a transformation of humanity's relationship to nature (from an exploitative "will to will" to an enabling releasement) and of our understanding of nature itself; it is also cast as the overcoming of nihilism. In this section, I first trace the beginnings of his thinking about technology in *Contributions to Philosophy (From Enowning)* and *Mindfulness*, written between 1936 and 1939 and, second, sketch the contours of Heidegger's *The Question Concerning Technology* and highlight its kinship with *Being and Time*.

1. Contributions to Philosophy (from Enowning) *and* Mindfulness

In these two texts, we see the rise of what Heidgger calls "being-historical thinking," or what he will later refer to as the "history of being," the dominant frame of his later philosophy. And a major part of this frame is the connections he makes between technology, nihilism, and nature.

In *Contributions*, Heidegger introduces a concept that would become a precursor to the notion of *Gestell*: "machination," which he deems "the domination of making and what is made."[81] Heidegger is adamant that his notion of "machination" does not denote the popular sense of that term, which is mainly pejorative. Machination, in other words, does not refer merely or primarily to when a person deceives, schemes, plots, and so on. In much the same way that inauthenticity from *Being and Time* is not a term of moral evaluation, but rather an essential part of Dasein, machination, in the context of Heidegger's "being-historical thinking," is an aspect of the "sway of being":

> In the context of the being-question, this word does not name a human comportment but a manner of the essential swaying of being. . . . Rather, the name should immediately point to making, which we of course recognize as a human comportment. However, this comportment itself is only possible on the basis of an interpretation of beings which brings their makeability to the fore so much so that beingness is determined precisely as constancy and presence.[82]

Machination, in other words, is a key ingredient in what Heidegger will later call the history of being. It is the alpha and omega of the history of metaphysics: "[I]t is very difficult to grasp historically the emergence of what is machinationally ownmost to beings, because basically it has been effectively in operation since the first beginning of Western thinking."[83] In *Mindfulness*, he is even more explicit: "'[M]etaphysics' . . . always means domination of being that is determined by thinking as representing."[84] The metaphysical roots of modern science, particularly the notions of causality and explanation, derive from Christianity's notion of being as created, but this notion itself derives, in Heidegger's view, from a mistaken interpretation of key concepts in Aristotle, especially *physis*.

Heidegger sees machination as part of what he calls the "abandonment of being," which involves a number of components, two of which concern us here: "calculation" and the "divesting, publicizing, and vulgarizing of all attunement."[85] Calculation, he says, "comes to power primarily by the machination of technicity, is grounded in terms of knowing the mathematical."[86] It is a kind of comportment that crowds out all alternatives and impoverishes our experience. Indeed, Heidegger sees this as the rise of "lived experience," a condition in which we frantically and futilely chase after experiences, adopting readymade, artificial modes of identity, fascinated by current culture, and passively absorbing its language: "The word then is only the shell and magnified stimulation, in which there can no longer be a connection to a 'meaning,' because all gathering of a possible mindfulness is removed and mindfulness itself is scorned as something strange and weak."[87] But keep in mind that in Heidegger's view, this is not a purely modern condition; it merely reaches its apex in modernity: "[B]ecause since long ago man, and modern man in particular, calculates everything (and even being) according to power and powerlessness, usefulness and advantage, success and uselessness, he is not capable of hearing any word of be-ing and of thinking its truth without initiating his calculation."[88] We have seen these themes before, and will return to them in greater detail in future chapters.

2. "The Question concerning Technology"

Heidegger begins his inquiry in this essay by seizing upon the customary view of technology and what people uncritically assume to be the "essence" of technology. He calls this the "instrumental and anthropological" definition of technology, which holds that it is "a means to an end" and "a human activity": "[T]o posit ends and procure and utilize the means to them is a human activity. The manufacture and utilization of equipment, tools, and machines, the manufactured and used things themselves, and the needs and ends that they serve, all belong to what technology is."[89] Just as he does in *Being and Time*, Heidegger here probes the "self-evident" surface of the reigning conception of the phenomenon to see what it is actually founded on, what hidden meanings it harbors. He questions the nature of instrumentality by pointing out that it implies the production of effects, and hence it rests on the notion of causality. After a discussion of the roots of our thinking about causality in Aristotle, Heidegger suggests that what unifies the different modes of causality and grounds them in a singular source is "presencing," the sheer arising and enduring of a thing. This occurs only through a "bringing-forth" (*poiesis*), which can be natural (*physis*) or artificial (*techne*): "*Physis* also, the arising of something from out of itself, is a bringing-forth, *poiesis*. *Physis* is indeed *poiesis* in the highest sense. . . . In contrast, what is brought forth by the artisan or the artist . . . has the bursting open belonging to bringing-forth not in itself, but in another, in the craftsman or artist."[90] This is important because Heidegger's understanding of the poetic and what we normally term the aesthetic includes and embraces the natural, instead of being opposed to the natural; indeed, the natural is the poetic par excellence. The way of revealing in technology, however, is not a bringing-forth: "the revealing that holds sway throughout modern technology does not unfold into a bringing-forth in the sense of *poiesis*. The revealing that rules in modern technology is a challenging, which puts to nature the unreasonable demand that it supply energy that can be extracted and stored as such."[91] Technology is a kind of attack on nature whereby the latter is transformed, via the mathematization of nature in modern physics, into a "gigantic gasoline station." Thus every merely present being is referred to its use as energy. The modes of being referred to in *Being and Time* as the present- and ready-to-hand are now seen as deeply interwoven: while the ready-to-hand is explicitly referred to an immediate use, the present-at-hand is implicitly referred to an eventual use; that is, it is studied theoretically only so that it can be put to some future use.

Taking all of these ideas together, Heidegger claims that the essence of modern technology is "enframing" (*Gestell*): "Enframing means the gathering together of that setting-upon which sets upon man, i.e., challenges him forth, to reveal the real, in the mode of ordering, as standing-reserve."[92] When Heidegger says that the essence of technology is nothing technological, and that it is not a mere means, he is basically saying that it is nothing ontic (it does not refer to entities within or facts about the world) but rather ontological (it refers to a way of revealing and a mode of being). *Gestell* is the way that beings show up in the modern age. It is the dominant world-horizon that determines what kind of sense things can have. It so thoroughly pervades our world, influences our behavior, frames our thinking, and skews our sight that we do not notice it; if we were fish, it would be the water.

The *Gestell* is essentially the totality of significance from *Being and Time*, with this decisive difference: its end is to prevent the possibility of breakdown. It is, in this sense, a form of "world domination" bent on the destruction of the earth because it seeks to overcome the struggle between the two poles once and for all, to force nature to reveal all of her secrets. Heidegger sees this phenomenon exemplified in the advent of atomic energy. But he insists that the real danger of technology is not its military as opposed to its peaceful uses, but rather its peculiar way of revealing. The contradiction in this project is that to be a world means to be finite. Its cohesion and meaning are based on there being an end—a final "in-order-to—and, as the existential analytic reveals, this end is the finitude of Dasein, its being-toward-death. The regime of technology, then, is for Heidegger a supreme illusion and an elaborate denial of death, the epitome of what in *Being and Time* he called "fallenness," and it is not even really a world because it does not have *bounds*: it motors on to the tune of infinite progress. It does not acknowledge the finitude of human life.

The danger of this project is twofold. First, by liquidating everything into standing reserve, man "comes to the very brink of a precipitous fall; that is, he comes to the point where he himself will have to be taken as standing-reserve."[93] We recognize this notion in phrases such as "human resources" and "human capital." The irony is that this is just when man comes to see himself as "lord of the earth." As C. S. Lewis put it in *The Abolition of Man*, "man's conquest of nature turns out, in the moment of its consummation, to be nature's conquest of man."[94] The greater man's power over nature, the further he falls from his own nature. Heidegger says this idea gives rise to another illusion, that "man everywhere and only encounters himself."[95] On the contrary: since man has forgotten his essential connection to and dependence on being/*physis*, he has forgotten his own essence. This is the basis of Heidegger's critique of humanism and ties into the second reason *Gestell* presents a danger: it "drives out every other possibility of revealing.... Where *Gestell* holds sway, regulating and securing of the standing-reserve mark all revealing. They [humans] no longer even let their own fundamental characteristic appear, namely, this revealing as such."[96] The technological mode of revealing is distinct and dangerous because it severely restricts the clearing through which beings can become manifest. It is a one-dimensional world.

This process erodes the essence of human freedom: the ability to disclose being and to recognize the ontological difference between being and beings. So technology masks Dasein from itself, which is to say, using the language of *Being and Time*, that it drags Dasein into the state of fallenness and inauthenticity. All these terms really mean is that Dasein tends to interpret itself in terms of and lose itself in entities within the world. Since technology pervades every sphere of life, it is harder to find opportunities for the breakdown and breakthrough required for authenticity, to glimpse ways of revealing and possibilities of being outside of *das Man*.

Haar connects this with what he sees as an oversight in *Being and Time*. When the referential totality is disrupted and tools show up as merely present-at-hand, the theoretical attitude of disinterested study and contemplation emerges, a posture wholly unconcerned with objects' use. In his later views on technology, however, Heidegger comes to believe that this posture is itself a part of the project of mastering nature aimed at gaining complete control over the object. Young echoes this sentiment: "[F]ar from being concerned to disclose the world in its 'ownness,' science is just another disclosure of it in a work-suitable way, another disclosure of it as resource. Natural science, therefore, is not an alternative to the technological disclosure of b/Being [sic] but a part of it."[97] This is why Heidegger claims that technology makes science possible: the step back from practical engagement with the object is only a means toward controlling it more effectively. Science always already contains a reference toward its eventual application and use.

Like Dreyfus, Haar seems to think that Heidegger subscribed wholeheartedly to an instrumentalist view of nature in *Being and Time*. However, as I discussed in a previous chapter, Foltz has convincingly shown that Heidegger's decision to start with the standpoint of average everydayness—essentially, the "work-world" (*Werkwelt*)—is strategic and methodological: his aim is to identify and peel away the derivative senses of nature correlated with certain modes of intentionality in order to arrive at the original phenomenon of nature. While the latter is not sufficiently worked out in that text, there are scant but significant signs that Heidegger saw the need to develop a notion of "authentic use," a more appropriate way of engaging with things after the "dark night of the soul" of anxiety and resolution; this third way—not treating things as ready-to-hand or as present-at-hand—would be correlated with a third sense of nature. And it is on precisely this note that he ends the essay on technology. Citing a line from Holderlin, "But where danger is, grows also the saving power," Heidegger claims that *Gestell* is inherently ambiguous and may actually contain the seeds of its own overcoming.[98] That is, technology stretches us to a kind of breaking point whereby we can first come to see our proper relationship to being.

Technology is not a mere human artifice, but one of the ways that Being shows itself to us. This opens up the possibility of a turn toward a new epoch that accommodates human freedom and does not exploit nature. The cultivation of a healthier relationship to technology involves rediscovering the ancient proximity of *techne* and *poiesis*, technology and art. This brings us full circle to the first section of the chapter: the work of art as the occasion for discovering the play between world and earth. Art serves as a countermeasure against the one-dimensionality of technology. The solution to the crisis of modern technology is not more or less technology, but the cultivation of a different attitude toward and understanding of it. For Heidegger, this consists in the standpoint of "releasement" and "poetic dwelling."

IV. Poetic Dwelling

There are four concepts that fall under this general rubric: poetic dwelling, the thing, the fourfold, and *Gelassenheit*. Together, they constitute the closest Heidegger comes to setting forth an environmental ethic, or at least the foundation for one. In this section I sketch the contours of these concepts.

In "Building Dwelling Thinking," Heidegger defines the human essence as "dwelling" and relates this to the notion of "caring-for": "Real caring-for is something positive and happens when we leave something beforehand in its nature, when we gather something back into its nature, when we 'free' it in the real sense of the word into a preserve of peace. . . . The fundamental character of dwelling is this caring-for."[99] There is an obvious connection here to *Being and Time*, where the structure of being-in-the-world is defined as care. Death denial, preoccupation with beings, and submersion in *das Man* all involve a failure to properly relate to oneself and one's own time. Authenticity does not consist in an ascetic withdrawal from the world, but a different disposition toward it; just so, poetic dwelling is not a call to Luddism or quietism, but to understand and thereby actualize the human essence and free the being of nature.

Poetic dwelling comprises a cluster of concepts Heidegger treats throughout his career, not just in its later phase: truth, freedom, *poiesis*, care and authenticity. The essence of Dasein is its existence, which is to say its being outside of itself and open to the world, a condition it can deny or embrace, and in this capacity for disclosure lies its gift, the ability to allow things to emerge and unfold in their own way. In a sense, humanity does not have an essence or nature, since its distinctiveness lies in its special connection with being/*physis* as the finite, open-ended clearing through which beings can become manifest in myriad ways. It can dwell *poetically* because it is a clearing of being; thus detached from beings, it allows them to come forth (*poiesis*). It can poetically *dwell* because it is finite: it is bound to a particular historical situation and destined for death. Here again, we can match these terms with their correlates in *Being and Time*: the poetic corresponds to Dasein's ecstatic futurity, its capacity to transcend present entities and project itself upon possibilities handed down from its past, and dwelling corresponds to its facticity, its always already finding itself grounded in and limited by a particular historical situation.[100] In technological thinking, Dasein tries to completely elude its facticity, to uproot itself from its temporal-historical finitude, to escape the entropy of the earth, to, in effect, conquer death. The poetic posture ceases to see facticity as a curse and a limitation to be negated and encounters it instead as a gift and a possibility to be cultivated. Poetic dwelling points to the balance and integration of these two poles of facticity and transcendence.

Poetic dwelling encounters entities as "things," not as "objects." Most of the time, we encounter things only to the extent that they fulfill a function as tools in our environment or as bare objects occupying space. In later texts such as "The Thing," however, Heidegger puts forth another interpretation. The classic example of the hammer used to illustrate "tool-being" in *Being and Time* finds its counterpart in the example of the jug used to illustrate "the thing" in the eponymous essay. Here we see recapitulated the same threefold analysis Heidegger applied to the work of art, noting that it cannot properly be conceived as a substratum, a sense-manifold (*aistheton*), or a form-matter composite, because all of these treat it merely in its being for humans and not in its independent and self-contained nature. As he says,

> no representation of what is present, in the sense of what stands forth and of what stands over against as an object, ever reaches to the thing *qua* thing. The jug's thingness resides in its being *qua* vessel. We become aware of the vessel's holding nature when we fill the jug. . . . When we fill the jug, the pouring that fills it flows into the empty jug. The emptiness, the void, is what does the vessel's holding. The empty space, this nothing of the jug, is what the jug is as a holding vessel.[101]

We can note two things here. First, the essence of the thing is encountered only in its use, but this kind of use is distinct from both the "using up" of technological thinking and the interpretation of the thing as a tool within a totality of significance. Second, the void or nothing of the thing refers to the *physis* that brings itself forth and can be shaped anew through *techne*. In other words, by encountering things in this authentic way, humans attend to things' proper possibilities, becoming a kind of conduit for nature to creatively emerge in and through the thing. In this sense, technology can be seen as tutoring rather than trapping nature. Heidegger elaborates on this idea, insisting that it is not merely an issue of the potter imposing his idea of the jug on a recalcitrant matter:

> [The potter] shapes the void. For it, in it, and out of it, he forms the clay into the form. From start to finish the potter takes hold of the impalpable void and brings it forth as the container in the shape of a containing vessel. The jug's void determines all the handling in the process of making the vessel. The vessel's thingness does not lie at all in the material of which it consists, but in the void that holds.[102]

In the discussion of the thing, then, we find the constellation of a series of concepts bearing upon the third sense of nature found in *Being and Time*: *Gestell*, the destruction of the earth, nihilism, and *poiesis*. The work of art is presented as the proper way to think of things in general, things are put forth as a more appropriate way to think of the being of entities, the latter is conceived of as nature/*physis*, and the cause and overcoming of nihilism are correlated with the revival of a deeper sense of nature.

After the essay on art Heidegger's notion of world comes to take on a new meaning, and in the later 1940s he begins to refer to it as the "fourfold" of earth, sky, gods, and mortals. Many interpretations have been hazarded by scholars attempting to unravel the meaning of the fourfold, perhaps the most enigmatic of Heidegger's later leavings; it is easy to dismiss and hard to decipher. Here I want to draw on Julian Young's persuasive attempt to demystify the fourfold by relating it back to the existential analytic of *Being and Time* in order to show how it incorporates the concepts discussed above, as well as Heidegger's notions of *Gelassenheit* and authentic use. By showing that the fourfold is a more poetic way of expressing the structure of worldhood or referential totality of significance in *Being and Time*, Young's analysis lends credence to the claim that the later Heidegger remains anthropocentric.

Here is perhaps Heidegger's most extensive description of the fourfold:

> [E]arth is the serving bearer blossoming and fruiting, spreading out in rock and water, rising up in plant and animal. . . . The

> sky is the vaulting path of the sun, the course of the changing moon, the wandering glitter of the stars, the year's seasons and their changes. . . . The divinities are the beckoning messengers of the godhead. . . . The mortals are human beings. They are called mortals because they . . . are capable of death as death.[103]

Young provides a useful schematic for making sense of the fourfold by relating it to the existential analytic of *Being and Time* and makes the case that poetic dwelling in the fourfold is Heidegger's mature, integrated view of humanity's place in the world. As to the first two terms, earth and sky, which correspond to nature, Young follows Haar, Taminiaux, and others in noting that nothing like these show up in *Being and Time*'s notion of care rooted in the threefold structure of temporality and suggests, "Had Heidegger paid more attention to [the third sense of nature], rather than simply recording its existence and passing on, the turn to the later thinking might have occurred much sooner."[104] Heidegger appears to lump all of the rich diversity of the animal, vegetal, and inorganic realms under the rubric of "earth," and this is thought by some of his environmental interpreters to indicate some sort of nonanthropocentric shift. Yet in this same essay he reiterates the sharp ontological distinction, made elsewhere, between humans and animals, emphasizing that one of the great mistakes in the metaphysical tradition was the characterization of the human being as an animal "plus" something else.

Heidegger occasionally characterizes the modern technological epoch as the "loss of gods," which is to say that we experience the gods *as* lost, not that more and more people cease believing in the existence of God. It means that we have lost the power to take over a creative relationship to our past as heritage, to repeat the possibilities handed down to us for our own time in an original way. Humans have ceded this power precisely by arrogating the power of nature to themselves and ignoring their natural and existential limitations; to borrow Charles Taylor's phrasings, humans have adopted an "exclusive humanism" that excludes an order that transcends them by setting up a this-worldly, self-sufficient, "immanent frame."[105] Ironically, the attempt to exclude "supernature" (humanism), which coincided with the conquest of nature and, inadvertently, human nature (*Gestell*), led to the view that nature was all there is (naturalism), and this is what led to the search for a new understanding of nature.[106] In other words, the rise of naturalism caused a backlash, first with the romantics in the nineteenth century, and later with environmental thinkers and activists in the twentieth, that interpreted nihilism as a consequence of naturalism and sought a more unified or holistic vision of nature as a panacea.

The lesson to be drawn from this parallel between the early and later work is that the supposed contrast between the anthropocentric-existentialist Heidegger of *Being and Time* and the nonanthropocentric-pantheist later Heidegger is overstated; the main difference pivots on his position on nature. The fourfold, then, is a poetic way of articulating a vision of the world that restores a proper balance between nature and humanity. In this sense, the fourfold can be plausibly seen as the integration of the earlier focus on human Dasein and the later focus on nature. It also implies an ought: not just what the structure of the world is, but how we should relate to it. If the fourfold points to that which is to be cared *for*, poetic dwelling refers to the "how" of the caring. This brings us to the final piece in the puzzle of poetic dwelling: *Gelassenheit*.

In the later works, the anticipatory resoluteness of *Being and Time* gives way to "releasement" (*Gelassenheit*), a notion Heidegger appropriated from Meister Eckhart, as the mature disposition toward being in which Dasein overcomes its willfulness and tendency to impose its categories of understanding on things. *Gelassenheit* is thus the countermeasure to *Gestell*. Heidegger also renders it as "meditative thinking" as opposed to "calculative thinking." Bret Davis, author of the sole monograph on Heidegger's treatment of the will, in perhaps the most extensive analysis of *Gelassenheit* in Heidegger's corpus, concludes that the tension between the will and its releasement pervades all but the entirety of Heidegger's thought and that the difference between early and later Heidegger on the will is more a matter of yin and yang than black and white.[107] Again, the difference in the style of presentation in the later works—such as dialogues and poetic musings—should not be mistaken for some radical shift in outlook: meditative thinking/*Gelassenheit* and calculative thinking/willfulness are close cousins of authenticity/inauthenticity. Just as authenticity does not mean a withdrawal from all worldly involvements, but appropriating the possibilities available to us in a conscious and creative way, *Gelassenheit* is not a posture of inaction, but is actually more rigorous, disciplined, and demanding than its counterpart. It counsels not the flight from Dasein's fallen condition, but equanimity in the midst of the world: "Meditative thinking demands of us not to cling one-sidedly to a single idea, nor to run down a one-track course of ideas. Meditative thinking demands of us that we engage ourselves with what at first sight does not go together at all."[108] This is contrasted with calculative thinking, the way of *Gestell*, which "never stops, never collects itself," and only discloses things in one way. It is imperative, however, to note that Heidegger does not call for the abandonment of calculative thinking for meditative thinking: "each [is] justified and needed in its own way."[109] *Gelassenheit* involves a simultaneous "yes" and "no" towards technology, criticizing its excesses but recognizing that it is mysterious. We do not fully understand its meaning, and it may actually harbor seeds of renewal.

Conclusion

I have attempted to show that the shift from Heidegger's early to later work is not from anthropocentrism to non-anthropocentrism, or from willfulness to *Gelassenheit*, or from an instrumentalist to a romantic view of nature. Heidegger's view *is* nonanthropocentric in the sense that he is trying to curtail the interpretation of being as presence and its current form of technological thinking (which is, at the same time, a critique of humanism) in order to open a space in which being can freely manifest itself. But at least two things speak against calling him a nonanthropocentrist. First, this is not the kind of nonanthropocentrism that some environmental philosophers have in mind. The ecocentric land ethic advocated by Leopold and Callicott, for example, is based on naturalistic sources such as Hume, Darwin, and scientific ecology, which tend to regard humans as just another animal species and as parts of the whole of nature. Second, his thought is anthropocentric ontologically because in the later work, as in *Being and Time*, Heidegger does not integrate human being with animal, vital, or natural being. The early critics of Heidegger I discussed in chapter 1 had it mostly right. I concur with Jonas's objection that, as Zimmerman notes, "what Heidegger really objected to was placing humans in any natural scale. Though condemning the technological domination of nature, Heidegger was never a 'bio-centrist,' but rather a Gnostic who viewed humans as aliens adrift in an indifferent or even hostile cosmos."[110] Löwith shares a similar concern, claiming that by setting out from human consciousness, Heidegger never arrives at the vision of a cosmos to which humans conform. On his reading, Heidegger provides us with a secularized form of the Christian model of stewardship for creation, yet without the "man at the pinnacle of creation" cosmology that underwrote it.[111]

Moreover, Heidegger's aversion to value thinking prevents him from deriving any ethical norms from the nature of things. From an ethical perspective, he ends up stuck, ironically, in the same place as the scientific naturalist trying to derive ethical "oughts" from ecological "is's." As is common among "holistic" environmental activists, writers, and thinkers, it is just assumed that the person who is mindful of Nature/Being will do the right thing. But just as the "call of conscience" from *Being and Time* is devoid of any specific content, so the "message of the gods" in the fourfold of the later work is contingent on the sending of Being. The vision of the fourfold *sounds* holistic, but the ontological gulf between humans and animals is maintained, and the primary focus remains the relation between humanity and Being. A grand narrative is spun that, though situating and subordinating humans within a world-historical process of the revelation of Being, still manages to make them the lead role as those on whom being depends for its manifestation and on whom beings appear to depend in order to be fully disclosed.

While Heidegger's prescription of poetic dwelling is suggestive for an environmental ethic, I think it is too vague and subtly anthropocentric. This subtle anthropocentrism is exemplified in Heidegger's notions of the thing and the fourfold. The concession of some kind of interiority, horizon of sense, or world to nonhuman beings (customarily referred to as the "reenchantment of the world") is an important move in an environmental philosophy, but we must be careful in how we do so, and do so in a way that is consistent with modern science. A jaguar "gathers" a world in a very, very different way than a jug does. The latter is a heap, not a whole; an artifact, not an organism. To haphazardly conflate the two as things, and to proclaim that things gather a world, conceals in its poetic obscurity a human projection that overlooks important differences that we can recognize and maintain without somehow violating the irreducible singularity of the beings in question. Heidegger's approach appears at times to border on an "animism of things"—and animism is anthropocentric! By disqualifying all of the traditional rational and evaluative ways of classifying the natural world, Heidegger is forced to describe nature in a poetic way that bears no discernible relation to the nature discussed by natural scientists. In short, he may not be using philosophical categories and theoretical abstractions to speak about nature, but he is using poetic devices, and these are still human.

Heidegger flirted with an alternate approach in his 1929–30 lectures, where he tried to integrate Uexküll's pioneering work on animal environments with his fundamental ontology and analysis of world. Despite the promise of this path, he spurned it in favor of a more poetic disclosure of nature that, while laudably articulating a positive alternative to the productive and objective senses of natural things, ends up vague in content, unacceptably hostile to, neglectful of, and incommensurable with the sciences, and ultimately unhelpful in forging a viable environmental philosophy and ethic. In the following chapters, I attempt to recover Heidegger's early approach by supplementing it with Nietzsche's own investigations into life and nature. But first, we need to look at Heidegger's fraught relationship to Nietzsche's philosophy.

6

Nature and Nihilism

Heidegger's Confrontation with Nietzsche

Picking up the trail of nihilism sketched in earlier chapters, here I examine Heidegger's view of nihilism as the logic of Western metaphysics, link it to his understanding of humanity's relation to nature, and compare it with Nietzsche's view of nihilism. At first, he sees Nietzsche's thought as the antidote to nihilism because of the latter's attacks on traditional metaphysics, idealism, and scientific naturalism, as well as Nietzsche's understanding of living being. Later on, however, Heidegger concludes that Nietzsche's thought—especially its call to naturalize and re-animalize human being—is the very essence of nihilism and must be overcome. Moreover, in opposing Nietzsche, Heidegger embraces an antinaturalist position at odds with the theoretical biology of his earlier thought.

I argue for the following cluster of claims. First, nihilism is one of the principal concerns of Heidegger's mature thought; his approach to the question of being aims to draw the problem of nothingness or nihilism into ontology proper. Second, his treatment of nihilism parallels his treatment of nature; while they are not explicitly engaged in his earlier thought, they come to play a prominent role in his middle and later thought. Third, the skeleton key to Heidegger's account of nihilism is his confrontation with Nietzsche's philosophy in the late 1930s; this is perhaps Heidegger's most important philosophical encounter. Fourth, Heidegger's interpretation of Nietzsche is mistaken because: a) in place of the more transcendental approach of fundamental ontology in *Being and Time*, Heidegger later adopts an implausible, idiosyncratic view of the "history of being" inspired largely by his own intellectual and cultural milieu; b) Heidegger wrongly interprets Nietzsche's views on the will to power, psychology, and value-theory anthropocentrically. Nietzsche's account of nihilism is genealogical, not merely ontological, and has what can, with qualification, be called "naturalistic" roots.

I proceed as follows. First, I sketch Heidegger's notion of the "history of being." As a number of commentators have pointed out, this became the regnant *Gestalt* of Heidegger's later philosophy, displacing the project of fundamental ontology found in *Being and Time*. Here I make three points. First, the turn to the history of being is by no means an abrupt shift, but should be seen as the attempted fulfillment of the unfinished segments of *Being and Time*, the "destruction of the history of ontology." Second, as scholars such as John Caputo, Charles Bambach, and Michael E. Zimmerman have pointed out, the operative view of *historical decline* in Heidegger's history of being is by no means original; it is reflective of a pervasive mood in post–World War I German intellectual culture captured by thinkers such as Alfred Baumler, Ernst Jünger, and Oswald Spengler. Heidegger's view of modernity as a nihilistic, technological wasteland was very much par for the cultural course and has important and disturbing similarities with national socialist–related thinkers. Third, the history of being is the context in which we must approach Heidegger's interpretation of Nietzsche because his ultimate view places Nietzsche as the culmination of Western metaphysics, that is, nihilism.

In the second section, I argue that Heidegger's epic survey of Nietzsche's thought, which unfolded from 1936 to 1942, particularly in the fourth volume, *Nihilism*, is the key to his account of nihilism. I begin by unpacking the mainsprings of this account. In the first of the Nietzsche lectures, Heidegger is convinced that Nietzsche pointed the way out of nihilism and interprets Nietzsche's philosophy of art as basically consistent with his own fundamental ontology in *Being and Time*.[1] In the final Nietzsche lecture, however, Heidegger changes his mind: Nietzsche is now portrayed as the consummate nihilist, the fulfillment of a process begun by the Greeks' interpretation of being as presence, leading to the rise of subjectivity and the "will to representation" in Descartes and German idealism, and coming to fruition in the philosophy of the "will to power," which motivates the "will to will" of technology, the *Gestell*. This change in perspective, however, is not merely about Nietzsche's thought: it is also the hinge for his turn away from the willful strains of his earlier thought toward *Gelassenheit*, from fundamental ontology toward the history of being, and toward his later notions of the thing and the fourfold—toward, in other words, his mature philosophy of nature that I detailed in the previous chapter.

Third, I isolate the chief problems in Heidegger's interpretation of Nietzsche and introduce Nietzsche's positive views. When all is said and done, Heidegger's ultimate view can be nothing other than a willful misrepresentation in order to fit Nietzsche into his neat narrative of the history of being. He thinks that Nietzsche's "metaphysics" of the will to power is actually the projection of a deficient mode of human comportment onto all beings that attempts to humanize everything. By reducing being to value, Nietzsche purportedly transforms everything into a projection of the human will. Yet Heidegger ignores many subtleties and key distinctions in Nietzsche's views. My contention, to be fleshed out in the final chapters, is that the very Nietzschean notions Heidegger ignores or misrepresents—will to power, psychology, values, and the nature of nihilism—actually point to problems and lacunae in his own view, and point back to the more naturalistic themes in his earlier work on Aristotle and Uexküll.

I. Nihilism in the Later Heidegger

In this section, I lay out Heidegger's notion of the history of being as it gestates in his earlier work and flowers in his 1935 *Introduction to Metaphysics*. This will carve out the context in which we have to see his encounter with Nietzsche.

1. The History of Being: Metaphysics as Nihilism

After *Being and Time*, Heidegger begins to talk more about the specific content of constituted worlds—the ancient, medieval, modern, and technological—rather than just the a priori, formal, ahistorical structure of Dasein's understanding of being. As such, he breaks with the phenomenological method and engages in a kind of speculative philosophy of history. Michel Haar explains how the latter became the dominant framework in Heidegger's later thought: "[T]he history of being becomes the guiding thread and the primary condition for all phenomenology. The world in *Being and Time* was the horizon sketched by the network of references effected by equipmental beings; it was ahistorical."[2] Fundamental ontology results, paradoxically, in the conclusion that ontology—and the method for doing it, phenomenology—is not fundamental. It is made possible by being's historical process of manifestation and withdrawal.

An even further consequence of this, however, is that Heidegger's own transcendental account in *Being and Time* is swept up in the tide of historical contingency. He realizes that the starting point for his own analysis—Dasein's everydayness, which is dominated by a technological understanding of beings as *Zuhandenheit*—is determined by the *Gestell* of modern technology, an ontological frame sent by being. Despite the increased power attributed to being's historical nature, Heidegger will persist in the attempt to retain some ahistorical aspects of Dasein and world disclosure—*Gelassenheit* and the fourfold, respectively, which are united in the notion of *Ereignis*—that underlie *all* epochs of being. So the tension between a "transcendental Heidegger" and a "being-historical Heidegger" does not entirely abate after *Being and Time*, but persists through the end of his work.

The history of being, then, is an account of the major ways in which being has shown itself in various "epochs" since its apparent emergence in the Greeks. The content of an epoch is the way being shows up "proximally and for the most part" in a particular world horizon; it is not the only way being can show up, but it is the dominant way, the "default" ontological setting. But epoch does not just refer to the mode in which being presents itself; it also refers to being's withdrawal. As Haar notes, "Every epoch of History is *epoche*, which means 'holding itself back,' 'self-suspension,' or '"withdrawal' of being, which goes hand in hand with manifestation."[3] Each epoch also has in common the "nothing" as an ahistorical aspect that underpins it; if the epochs are the beads, the nothing is the necklace. As I discussed in chapter 5, Haar makes the compelling case that this ahistorical dimension is what motivates Heidegger to introduce the notions of earth and *physis* in the 1930s.[4]

Heidegger begins to refer to humans' experience of the nothing with the term *Ereignis*. As Haar explains,

> *Ereignis* . . . refers to the other side, the nonhistorical side of the "historical" reign of Technology. *Ereignis*, which does not belong to the History of Being, brings about with the abruptness of lightning the simple thought "outside any epoch" . . . [It] is the condition of entering into this non-metaphysical experience of the world that Heidegger sometimes describes as letting-be, or as the non-objectivized proximity of things, or as the completely non-anthropomorphic deployment of the four regions (earth/sky, mortals/gods) that reflect into each other.[5]

If the epochs are the "diachronic" dimension of Heidegger's later thought, then *Ereignis* is the "synchronic" dimension. Heidegger maps out four epochs—ancient, medieval, modern, and technological. There are elements of both continuity and discontinuity across the epochs. One thing they all have in common is the conception of being as presence. Roughly, for the ancients, to be was to be constantly present; for the medievals, to be was to be created by God; for the moderns, to be was to be represented by a subject; and for the age of technology, to be is to be "standing reserve" (*Bestand*) or raw material on hand for manipulation. The more important continuity, however, has to do with the trajectory of the epochs. As Haar observes, "The series of the epochs of Being obeys an inflexible and coherent necessity which, [Heidegger] writes, 'is like a law and logic.' "[6] And this quasilaw and -logic, Haar insists, has a "strong teleological structure." Haar makes a powerful case that, despite his criticisms of Hegel's philosophy of history, which is purportedly a prisoner of a subjectivist metaphysics, Heidegger's history of being is in many ways just as systematic and ambitious and that it resembles an "inversion of Hegelianism":

> [T]he Hegelian becoming of truth becomes the progressive establishment of the reign of errancy, the development of nihilism. . . . History is not the progress of consciousness toward self-transparency, or the absolute movement, but is the gradual loss of the sense of presence as clearing and withdrawal. The "evil telos" that orients History is the complete obscuring of the meaning of being.[7]

Technology is completed metaphysics; it is the end already contained in the Greek beginning. So Heidegger's view of history is somehow both progressive and declinist, married, as Caputo puts it, to an *arche* and an *eschaton*.

Both Haar and Bret Davis connect this progressive movement of the history of being with the rise of the "will to will," which wills for no reason, with no purpose or end in sight. We could refer to the "will to will" as a form of "historical nihilism." As Haar details, "Technology manufactures in order to manufacture, it exploits the earth in order to exploit it, stockpiles energy in order to stockpile it and not in order to respond to any 'actual' need. The doubling back of the will onto itself indicates its 'nihilism': it pursues no end, it develops onto itself to the point of the most complete irreality."[8] The will to will labors for the destruction of the will and the negation of the subject, since the human, too, becomes raw material for the preservation and enhancement of *Gestell*; the endgame is the disappearance of both subject and object. As Davis has it, "the fundamental (dis)attunement of will escalates until, abandoned to the pure immanence of the will to will, the will recognizes no other in its frenzied hunt for more control, more power, more will. The progressive emergence of the will in correlation to the increasing withdrawal of being thus provides a marked continuity to the history of metaphysics."[9] In other words, metaphysics is nihilism and was always about humanity's attempt to master and control nature (*physis*), but this only becomes clear and explicit in the age of technology. Somehow, however, Heidegger insists that being has always been in the driver's seat; it was not through human effort, but through being's own self-granting, that metaphysics commenced, humanism flourished, and the *Gestell* took hold. The fall, nihilism, is not Dasein's fault and cannot be reversed through Dasein's effort. To think that Dasein can will itself out of nihilism would be like trying to dig oneself out of a hole with a shovel.

2. Introduction to Metaphysics

Heidegger's concern to articulate the shape of the present only increases in his 1935 *Introduction to Metaphysics*,[10] where the rhetoric of nihilism swells further and the history of being begins to take shape. Since I have already showed how this text signals Heidegger's turn to *physis*, here I focus on three points in the text: the dramatic, martial, voluntaristic diction; the connection between nihilism and the corruption of language; and the narrative of historical decline. The text is a major turning point in Heidegger's thinking about nihilism because it portrays the collision of the two kinds of nihilism he identifies: historical and ontological. My view is that in this text, Heidegger tends toward the view that nihilism is not an ahistorical, ontological condition, but a historical one that man caused and can remedy. In *IM*, Nietzsche is cited nearly a dozen times and with almost unqualified reverence as the philosopher-prophet pointing to the overcoming of nihilism. This is tied in no small part to what Bret Davis has identified as Heidegger's "embrace of the will" in the mid-1930s.[11] I think Heidegger "tends toward" this position because, as is almost always the case in Heidegger's texts, we find ambivalence. For Davis, this ambivalence involves a vacillation between an embrace of the will and a turn toward *Gelassenheit*. I see a similar ambivalence over nihilism: in some texts, nihilism is portrayed as a condition humanity can overcome, while in others it is cast as part of the human condition and the nature of being itself.

First of all, the text is motored by a determination to fuse philosophy and the contemporary political situation: great thinkers come off as philosopher-kings who will deliver us from the slaughter bench of history. Heidegger paints a dramatic picture of the present, identifying the plight of the German people with that of the human essence. Germany is caught in a pincer between the two dominant ideologies of the day, capitalism and communism, which are, he claims, "metaphysically equivalent" because they depend on a neglect of the finitude of man and being and adhere to a progressive view of history ignorant of its metaphysical origins. The present is a moment of crisis characterized by the flight of the gods, the destruction of the earth, the darkening of the world, and so forth: "World is always world of the spirit. The animal has no world nor any environment (*Umwelt*). Darkening of the world means emasculation of the spirit, the disintegration, wasting away, repression, and misinterpretation of the spirit."[12] Cursing the darkness and overcoming nihilism demand a great struggle, the summoning of a collective will to turn around mankind's fallenness among beings and his forgetfulness of being. Heidegger employs a popular style of the time to capture this pathos: the martial rhetoric of struggle, will, power, creation, and destiny. Through some sort of titanic struggle with the retarding forces of modernity, the German people will seize and submit to their destiny (note the conflation of freedom and fate—true freedom is somehow both seizing nothing and submitting to nothing) and give birth to a new age free from the divisive dualisms of the past. The vaunted rhetoric at times flirts with parody. Consider a sample:

> [Germany] is the most metaphysical of nations. We are certain of this vocation, but our people will only be able to wrest a destiny from it if within itself it creates a resonance . . . and takes a creative view of its tradition. All this implies that this nation, as a historical nation, must move itself and thereby the history of the West beyond the center of their future "happening" and into the primordial realm of the powers of being.[13]

At this point Heidegger appears convinced that the world situation can be turned around through human ingenuity: if Dasein is strong enough, the ontological recession can be beaten back. Heidegger will later recant this view, since he realizes that it recapitulates the modern progressivism that he thinks is part of the problem. This pushes him to define Dasein's role down, making it more subject to the slings and arrows of being's fortune, and to claim that the proper disposition is one of *Gelassenheit*. Yet this does not solve the underlying problem: that he has robbed Dasein of any means of differentiating between worthy and unworthy assignments, goals, and values by interpreting reason as intrinsically calculating and technical. It is not the power of discrimination among alternatives, but of openness, that will guide us; but this openness can accommodate any content. Striving to overturn the present order and submitting to it are equally legitimate responses to the current situation, and thus the difference between them becomes meaningless. As Stanley Rosen puts it, "The problem is that [Heidegger] states no alternatives in an explicit manner but leaves everything sufficiently vague as to justify either of two distinct inferences. . . . Very simply stated, openness to Being, or to that which regions, is compatible with doing nothing or with doing anything at all."[14] But in *IM*, the emphasis is undoubtedly placed on the "human side": of freedom, the leap, will, resolve, the struggle of the spirit, creation, and power. *IM* seems to hold, in other words, that Dasein can overcome nihilism, which is to say that nihilism is not a chronic condition, but a historical one.

Second, Heidegger ties the rise of nihilism to what he sees as the gradual corruption of Western language in the translation of Greek terms into Latin and in the grammatical modification of the word "being." Puzzled by the fact that in modernity the meaning of being is vague and indeterminate—*Being and Time* began, recall, by arguing that being has become the most universal, indeterminate, and abstract concept—Heidegger attempts an etymology of being. He thinks that the decreased power of Western languages to respond to being is a symptom of the decline of a genuine experience of being and that this commenced with the codification of philosophical concepts or what we might call "prime words" in the Greek language and their subsequent translation into Latin. For the Greeks, being meant standing or enduring presence: "[T]his erect standing there, coming up, and enduring is what the Greeks understood by being. . . . Coming to stand accordingly means: to achieve a limit for itself, to limit itself. Consequently a fundamental characteristic of the essent is *to telos*, which means not aim or purpose but end."[15] Here the limit, end, or horizon of possibilities is not what constrains a thing's being, but what enables it to be in the first place. What Heidegger is pointing to here is a shift from a *determinate* sense of being grounded in experience to an *indeterminate* sense of being codified in a grammatical abstraction: the infinitive, "to be," which we take as the primary mode of being. He sees this as of a piece with a transformation in our understanding of language. So a shift in language is the expression of a shift in experience, but once the linguistic shift has occurred, it perpetuates the experiential shift that caused it. According to Heidegger, this dialectic of the progressive dissociation of language and being reaches a breaking point in the age of modern technology, in which we adopt a tacit nominalism: language is interpreted in an instrumental and anthropological sense. That is, language is merely a means toward ends projected by human beings, not nature or God or being itself. This "grammatical" interpretation of language—the view that language is a free creation of human beings and a kind of calculus or measuring tool—is one of the main causes of nihilism, since it ruptures the original connection between speech/language/thought and being, between *logos* and *physis*.

This would seem to imply that there is a correct way of speaking that can be excavated once the dross of idle talk and abstract concepts have been skimmed off. But since Heidegger insists that the entire edifice of Western rationalism is itself part of the dross, then the only purified language left is poetic language. Indeed, to demand a "correct" way of reasoning or thinking about being is to fall prey to a particular conception of truth that is equi-primordial with the Greek disclosure of being as standing presence. Since language is fundamentally poetic—a creative process that brings forth possibilities granted by being—authentic saying must be nonrational, lest it grow inattentive to the spontaneous emergence of being as *physis* and lapse into preoccupation with the beings that emerge, that is, positivism. In this text, then, we see Heidegger's vigilant distinction between the ontological and the ontic applied to a decline in language: there is pure, authentic, ontological, and upright logos and fallen, inauthentic, ontic, and *declined* logos.

Heidegger's approach to language here is a symptom of a deep tendency in his thought which John Caputo has labeled the "law of essentialization."[16] According to Caputo, Heidegger's thought embraces a "systematic valorization of *Wesen* over that of which it is the essence," or, put another way, it privileges the ontological or "deep" structure of things over the things themselves, the ontological over the ontic.[17] This logic is viral in his works: the essence of technology is nothing technological, the essence of language is not speaking, the essence of human being is nothing human, and so forth.[18] In each case, the essence leads back to being itself—to *physis*—and there is no room for meaningful distinctions about the things themselves. They are dissolved in the universal acid of the purification of language, which is to say that the distinction between the ontological and the ontic becomes suspect, and with it the power of language itself. This is exactly Rosen's definition of nihilism: the equation of speech and silence.[19] By insisting that extant philosophical language and rational terms are bankrupt, the philosopher must poetically create language anew in order to more adequately address being; but since he has forsaken intelligibility, he inures himself from criticism and dialogue, which is to say that his speech is equivalent to silence. Since philosophical language, like all language, is originally poetic, it has no purchase on nature and is supremely anthropomorphic. There is no categorial compass with which we may approach nature; it is simply a mysterious realm of things-in-themselves that slip the bounds of metaphysics and science.

The third important feature of *IM* is its embrace of a narrative of historical decline. In Heidegger's time, intellectuals were divided over the nature of history into roughly two camps: historical progressives and historical declinists. Zimmerman describes the progressives: "Affirming Germany's appropriation of Enlightenment cultural, political, and scientific values, neo-Kantians interpreted history as the gradual development both of more enlightened modes of social organization and cultural self-expression."[20] As we saw in *FCM*, Heidegger rejected the philosophy of culture typified by thinkers like Ernst Cassirer because it rests on a dubious dualism between man's natural being and his cultural or symbolic being. This intellectual paradigm shift, inspired by what Heidegger sees as a misinterpretation of Kant's metaphysical project, is what motors the modern shift away from ontology and toward anthropology and psychology and provides the confidence in the value and possibility of the gradual perfection of man's intellectual and moral powers and social and political situation.

The historical declinists fit into two camps. As Zimmerman explains, "Opposed to the progressive view of history were two groups: those who believed that history had no direction, and those who believed that history involved a decline from great beginnings."[21] Heidegger's view was substantially influenced by declinist thinkers such as Spengler and Jünger, two of Nietzsche's intellectual scions. Spengler arguably attempted to schematize Nietzsche's psychology of the will to power into a philosophy of history. As Zimmerman notes, "[According to Spengler], since human history lacks any overall meaning or purpose, each civilization views things from its own perspective, establishes its own table of values, and thus constitutes a unique type. . . . Decline sets in as this primal symbol loses its force."[22] Heroic, creative, charismatic individuals posit the values that define and orient a particular culture, but their values are not anchored in any eternal order such as cosmos, creation, or nature; they are successful because they supply the optimal conditions for the preservation and enhancement of individual and collective power.

In a similar vein, Heidegger attempted to expand his early fundamental ontology into a "history of being"—not an "empirical" history or an "interpretation" of actual events from a human perspective, but a "transcendental" history that laid out the a priori conceptual gestalts, epochs, or stages that being itself set up and in which humans played out their lives. Zimmerman spies such a connection between Spengler and Heidegger: "Something akin to Spengler's *Ur*-symbol is discernible in Heidegger's claim that each epoch of Western history (ancient, medieval, modern, technological) is governed by a particular mode of being that organizes all cultural practices and institutions."[23] Here, Kant's categories and Husserl's structures of transcendental subjectivity are transposed into the historical world horizons or epochs of being.

Now, despite these similarities with Spengler, Heidegger is critical of his approach in ways that will be important for his critique of Nietzsche. Zimmerman points out that "whereas Spengler regarded the will to power as the foundation for all cultures, Heidegger regarded it as the mode of metaphysical understanding characteristic of the near final phase of Western history, which is governed by the foundationless destiny of Being."[24] Heidegger will later claim that this is Nietzsche's fatal mistake: failing to see that his own understanding of being as the will to power and his transformation of metaphysics into psychology are historically contingent. Conceiving of the human spirit or transcendence as culture, and of culture as the creative positing of values, ignores humanity's dependence on being. Moreover, Heidegger believed that Western history had fallen from noble origins, whereas Spengler viewed historical gestalts as passing through a cycle of emergence, consolidation, and decline. Heidegger tends to see the modern age as a decline from the noble beginning in Greece in which the apprehension of being as *physis* briefly emerged but was subsequently obscured and repressed, whence the process of forgetting was compounded and eventuated in the modern technological age. Caputo has christened this tendency Heidegger's "mythology": "The privileged status of the early Greeks forms the core of a vast, overarching, and—it is now plain—highly dangerous metanarrative, a sweeping myth about Being's fabulous movements through Western history."[25] He notes that Heidegger violates his proviso in *Being and Time* that we must, above all, refrain from telling a story:

> The thrust of the argument in *Being and Time* is actually to discourage the mythologizing move, to discourage privileging any factual-historical interpretation of Being, and to concentrate on the formal structure of the understanding of Being. . . . Heidegger's turnabout on modernity and his privileging of a mythic age of early Greeks are central and defining features of the turn in his thought.[26]

Caputo cites Heidegger's changed view of modernity as key to his turn to mythology: "The whole of modernity is looked upon not as a period of breakthrough and discovery of the contribution of the subject (and hence of Dasein) [as it was, e.g., for Hegel] but as a subjectivizing of Dasein. Modernity is the age of the world picture."[27] And creating and clinging to "world pictures"—objectifications of spirit/life, worldviews, world-historical narratives like those Heidegger criticizes in *FCM*—are symptoms of nihilism. But as Caputo and others have shown, the history of being is just another "world picture." Heidegger is pulled away from a more modest, parsimonious approach and toward a more extreme, profligate, speculative one. Instead of describing Dasein's temporality and world "from within," he shifts to narrating the history of being "from without." He presumes, in other words, to be both in *medias res* and to be outside of his own time—outside of epochal determinations altogether. In the context of our discussion, Heidegger's lapse into mythology is a form of nihilism that is captured more by Nietzsche's notion of passive nihilism, in which humanity projects an imaginary, ideal world in the mythic past and hoped-for future in contrast to the present, fallen one, both yearning for the lost world and hankering for the next.

Moreover, Caputo adds that Heidegger excluded the Judeo-Christian traditions "in order to construct a native land and a mother tongue for Being and thought."[28] In this sense, Heidegger's thought strikes the pose of the "conservative revolutionary" that refracts the past and future through the present by invoking a golden past and promising a radically new, "other beginning" that is both a return and a revolution: we cannot articulate what the new alternative will look like and must simply have faith that the new dispensation will deliver us from the present evil.

In a similar vein, Lyotard has dubbed this "fetish" for the homeland "Heidegger's geophilosophy."[29] This should be a warning to those looking to "green" Heidegger. His mythologizing strain and attachment to a narrative of historical decline have much in common with green thinkers' and activists' affection for premodern peoples and epochs and their ambivalence toward modernity. The problem here is that "nature" is conceived *terrestrially* as earth, rather than *cosmically* as embracing the physical dimension that subtends the earth.

But the larger problem Caputo points to is Heidegger's attempt to fuse the ontological and the ontic. Once the "hermeneutic of factical life had given way to essential thinking, to thinking Being's own history," "the myth of a deep Essential Being, both structurally primordial and historically Greek, was firmly in place."[30] I dwell on this point because it highlights a paradox of greening Heidegger. If we are to use that side of his thought that is allegedly more nonanthropocentric and apparently more germane to environmental thought, we must sign on to his grandiose history of being, his law of essentializing, his rejection of modernity, his myth of the origin, his dismissal of evolution, and his rejection of a progressive reading of history. These are not just unsavory aspects that can be surgically removed from his later thought; they are vital parts in an intricate organism. If we retreat to his earlier work, however, we have to deal with what Caputo called his

Cartesianism: his ontological dissociation of human beings from animals and from nature. As we have seen with Löwith and Jonas, Heidegger's transcendental thinking uproots us from a cosmos, a nature in which we are at home, and relegates nature to a projection of human understanding, a correlate of consciousness. It is no accident that the third, poetical sense of nature gets short shrift in *Being and Time*; but when it does get worked out, it is within the context of a deeply prejudiced and ethnocentric view of history that has trouble underwriting a humanist ethic, let alone an environmental one.

Moreover, Bambach points out the naivete of ecological readings of Heidegger that praise his critique of modern science and technology:

> When we read the Nietzsche lectures as a critique of biologism, race, and the metaphysics of blood and consanguinity, we need to remember that Heidegger's rejection of these principles was grounded in what he deemed a more fundamental form of communal identity—namely, autochthony. . . . This critique of modern science as positivism has all too often been read in terms of an eco-poetic, deep-ecological critique of technological devastation and domination.[31]

Heidegger's later paeans to "letting things be" and "saving the earth," then, should not be seen as springing from genuine ecological concern, but as attempts to dilute and soften and generalize what in the 1930s was a militant political ideology: "Far from being a pastoral roundelay about the rural landscape," Bambach writes, "Heidegger's song of the earth in praise of rootedness and autochthony is part of a martial-political ideology of the chthonic that was deployed in the 1930s in the name of a German metaphysical-racial autochthony."[32] Heidegger rejects national socialism's positivism and naturalism because it "denies the essential historicity of a *Volk*." It is another form of nihilism, in other words, because it does not situate humans within nature. The jargon of autochthony, which is concerned with nature, and what Adorno called the jargon of authenticity, which culminates in the embrace of the will in the 1930s and thus with nihilism, are of a piece. Heidegger's move toward *Gelassenheit* and a poetical sense of nature is a reaction against his own nihilistic embrace of the will but is still riddled with attachments to troubling tendencies of his time.

Again, while Heidegger gives a different account of the source of the decline—he avoids either a racial/biological interpretation, *a la* national socialist ideologues—he nevertheless relies on a framework he fails to justify.

This framework is the skeleton of his history of being, the *Ur*-thought of his later work. His mature position will be that nihilism is an inescapable, existential-ontological condition that stems from the nature of being itself; the rise of the technological "will to will" is not a historical choice made by humans, not something within their control, but is rather an epoch or sending of being, the way that being has "decided" to show itself in the present age. Rather than try to overcome this condition, humans should (qualifiedly) embrace it. To paraphrase Ernest Becker, for the later Heidegger, the attempt to escape from (ontological) nihilism is the greatest cause of (historical) nihilism.

II. Heidegger's Nietzsche Interpretation

Heidegger's confrontation with Nietzsche is important for many reasons, but two stand out: it is both the most extensive and perhaps the most suspect interpretation of all his readings of the Western tradition. In regard to the first, Alan Schrift reports that "Heidegger published a greater volume of material on Nietzsche (over 1,200 pages devoted specifically to interpretations of Nietzsche) than any other figure in [the history of metaphysics]."[33] In regard to the second, Laurence Lampert marvels at how, in Heidegger's Nietzsche lectures, "Nietzsche appears as a metaphysician even though he scorned metaphysics, as a Platonist even though he thought of himself as the anti-Platonist, as a nihilist even though he said he overcame nihilism."[34] Walter Kaufmann pulls no punches:

> Heidegger's reading of Nietzsche rests on three clear and simple principles: first, we must discount and ignore Nietzsche's books. Secondly, Nietzsche's philosophy is to be found in his *Nachlass*; that is, in the notes that he himself did not publish and did not intend for publication. Thirdly, most of these notes can be ignored also, and the real philosophy of Nietzsche is to be found in the notes selected by Heidegger—almost exclusively from the *The Will to Power*. . . . Heidegger's approach to Nietzsche is a philological and methodological scandal that almost defies belief.[35]

Though the lectures were prepared and delivered between 1936 and 1942, they were not published in German until 1961, and as Charles Bambach has meticulously documented, they were marketed and received as Heidegger's second magnum opus after *Being and Time*.

While it is customary to separate the early and the later Heidegger, and thus to assume that Nietzsche only becomes an influence in the latter period, scholars such as David Farrell Krell and Jacques Taminiaux have shown that Nietzsche's thought not only was foundational for *Being and Time* but also led Heidegger to see the work's shortcomings. Krell supposes that Nietzsche can plausibly be regarded as the "regnant genius" of *Being and Time*, since his ideas underwrote Heidegger's analyses of mortality and temporality and supplied the notions of "the anthropomorphic base of metaphysical projections and the evanescence of Being as permanence of presence."[36] Taminiaux, meanwhile, has demonstrated that Heidegger's first lecture course on Nietzsche, *The Will to Power as Art*, not only endorses Nietzsche's conception of artistic creation as the "counter movement to nihilism" but likens Nietzsche's accounts of self-overcoming, the overman, self-transcendence or -enhancement, and rapture/ecstasy/intoxication (*Rausch*) to his own fundamental ontology; he insists that Nietzsche's major terms—will, affect, passion, drive—are not purely psychological categories, but should be seen as "essential moments of Dasein's constitution," like the existential categories of fundamental ontology.[37] Moreover, Taminiaux observes that Nietzsche's emphasis on the life-enhancing power of art was the "indispensable inspiration" for Heidegger's turn to *poiesis* in the mid-30s: "[The distinction between *techne* and *poiesis*] is lacking in the early analytic of Dasein, which seems to reduce the entire realm of *techne*, i.e., the technical know-how of the artisan or expert, as well as the art of the artist, to the level of practical circumspection enmeshed in everydayness."[38] As we saw in previous chapters, aesthetic experience is not addressed in the existential analytic; recall that this lacuna paralleled the absence of the third sense of nature. And it is thus unsurprising that, as Taminiaux discerns, when Heidegger analyzes Nietzsche's account of the lived body, he "even suggests that Nietzsche's notion of physiology not only has nothing to do with modern physiologism but stands in close attunement with the Presocratic understanding of *physis*."[39] In other words, Nietzsche's account of the connection between humans' creative self-transcendence and a new understanding of nature as a means of overcoming nihilism played a major role in the development of Heidegger's thought.

Indeed, we might see Nietzsche as the polestar around which Heidegger's thought revolves. If we follow Hannah Arendt in locating Heidegger's turn between the first and second volumes of the Nietzsche lectures, and endorse Bret Davis' recent claim that Heidegger's position on the will is the pivot point of his thought, then it looks as though Heidegger's final take on Nietzsche's thought is hardly a peripheral concern and is instead of ultimate importance for his thought as a whole. Heidegger's early and later interpretations of Nietzsche can be read as proxies for his early and later thought; and those positions, as it happens, have everything to do with the meaning of nihilism.

The picture that begins to emerge is that Heidegger's Nietzsche interpretation and his history of being—in other words, his account of nihilism—are riddled by biases native to his thought and rampant in his culture. Specifically, his dualism between ontological and historical nihilism, his aversion to value thinking (a reaction to the Neo-Kantian tradition), and his indebtedness to a number of national socialist-related thinkers (Baumler, Spengler, and Junger, to name a few) seriously skew his understanding of Nietzsche and nihilism. These add up to a kind of "perfect storm" that somehow led one of the 20th century's most brilliant thinkers to embrace such an implausible interpretation of the world situation. As we will see, Heidegger rejected what Dan Conway calls Nietzsche's genealogial and "naturalistic"[40] account of nihilism, which tried to trace the emergence of nihilism as a historical process dependent in part on biological and evolutionary factors, in favor of his own idiosyncratic view.

I proceed in this section as follows. First, I lay out the basics of Heidegger's interpretation from the fourth Nietzsche lecture: the nature of the will to power, the importance of values, Nietzsche's relation to Descartes and significance within modern philosophy, his call to "re-naturalize" or "animalize" humanity, and his misguided designs to overcome nihilism. Second, I rebut Heidegger's position by showing his misunderstanding of the will to power, his mischaracterization of Nietzsche's value theory as anthropocentric, and his neglect of the genealogy of nihilism.

According to Heidegger, Nietzsche's answer to the question "What is the being of beings?" is "Will to power." Invoking the traditional distinction between essence and existence, Heidegger claims that for Nietzsche, the will to power is the essence of all beings. Though he initially believed that Nietzsche's vision of humanity recognizing and owning up to its own will to power and creating new values was congruent with his own ideal of authenticity, Heidegger came to view the will to power as the metaphysical basis of the penultimate stage of nihilism. Nietzsche's vision, he thought, was "true" to the extent that it accurately depicted the modern world of the nineteenth century, but it was superficial because it was blind to the process of the history of being operating "behind its back." So there are two things to look at here: first, how Heidegger characterizes the will to power in itself, and second, how he integrates it into the arc of modern philosophy.

I think the most useful way to frame the first is Nietzsche's psychology: early on, Heidegger labors to clear Nietzsche of the charge of psychologism, but later, he becomes convinced that Nietzsche's thought is hopelessly entangled in a form of psychologism that reduces human beings to the status of animals and then projects this psychology onto beings as such. His critique of this psychologism is not on logical grounds, like Husserl's, but for ontological-historical reasons: Nietzsche was the inheritor of a modern metaphysical tradition that progressively interpreted being as will or subjectivity, and this itself is rooted in the older tradition of interpreting being as standing presence. The terminus of this trajectory is the reduction of being to will and that which is posited by the will—values. So Nietzsche's attempt to overcome nihilism by way of "renaturalizing" or "animalizing" humans is, Heidegger holds, the deepest entanglement of nihilism. The progress of the "overman" is really a regress to an "underman." Somehow, the thinker who is famous for his attack on the spiritually debilitating aspects of secular humanism becomes, in Heidegger's hands, the herald of humanism that appoints man the measure of all things.

Heidegger is emphatic that Nietzsche equates the will to power with being. As Lampert points out, "will to power, becoming, life, and being in the widest sense mean the same thing in Nietzsche's language."[41] There is a considerable debate over whether the will to power should be understood as a merely psychological principle or whether it is a metaphysical or "cosmological" principle. Nietzsche gave psychology tremendous importance: he deemed it the "doctrine of the development of the will to power" and insisted that it once again be seen as the "queen of the sciences" and as "the path to the fundamental problems," and it is the basis of his genealogical critiques of morality and religion.[42] It might therefore appear that he thought the will to power was merely an explanation for human activity and that we should not read him as a full-blown metaphysician. Lampert objects to this view, pointing out that there is "no positive evidence that Nietzsche abandoned the notion of will to power as metaphysics."[43] He argues that Kaufmann's psychological interpretation does not account for how persistent the metaphysical view of the will to power is in Nietzsche's thought and how it "provides a basis for his accounts of life and the activities of man."[44] So the will to power is not a psychological debunking of metaphysics, but, Lampert insists, "the application of one metaphysical view to other metaphysical views."[45] Heidegger agrees. For Nietzsche, he thinks, psychology becomes metaphysics:

> For Nietzsche, "psychology" is not the psychology being practiced already in his day, a psychology modeled on physics and coupled with physiology as scientific-experimental research into mental processes, in which sense perceptions and their bodily conditions

are posited, like chemical elements, as the basic constituents of such processes. . . . Nietzsche's psychology in no way restricts itself to man, but neither does it extend simply to plants and animals. . . . Nietzsche's psychology is simply coterminous with metaphysics. That metaphysics becomes a "psychology," albeit one in which the psychology of man has definite preeminence, lies grounded in the very essence of modern metaphysics.[46]

Crudely put, the essence or nature of all beings—including so-called "inorganic" beings such as molecules and atoms—is a kind of will or life force. As Ruth Irwin notes, "[Nietzsche's] theory of will to power attributes perspectives to other forms of existence that are outside the parameters of human comprehension. The concept is inorganic rather than being limited to breathing, living things. He regards each mode of being as having its own perspective and thus its own world."[47] A passage from *Beyond Good and Evil* illustrates this point:

> Suppose nothing else were "given" as real except our world of desires and passions, and we could not get down, or up, to any other "reality" besides the reality of our drives . . . for thinking is merely a relation of these drives to each other: is it not permitted to make the experiment and to question whether this "given" would not be sufficient for also understanding on the basis of this kind of thing the so-called mechanistic (or "material") world? I mean, not as a deception, as "mere appearance," an "idea" . . . but as holding the same rank of reality as our affect—as a more primitive form of the world of affects in which everything still lies contained in a powerful unity before it undergoes ramifications and developments in the organic process . . . as a pre-form of life.[48]

What we call "matter," then, is not essentially different from "life" or "mind," but an evolutionarily more primitive form of the same thing: the will to power. In response to the Cartesian problem of interaction between mind and matter, Nietzsche proposes to dissolve the problem by rejecting the view of the material world as purely extended, inert matter: "one has to risk the hypothesis whether will does not affect will wherever 'effects' are recognized—and whether all mechanical occurrences are not, insofar as a force is active in them, will force, effects of will."[49] He offers a unified vision of "the world viewed from inside, the world defined and determined according to its 'intelligible character.' "⁵⁰ Let me briefly note here that embracing will to power as a cosmic principle poses a problem for Nietzsche in the realm of biology: it suggests that he embraced vitalism, a view now widely discredited in biology. This is a serious obstacle to casting Nietzsche as offering a viable philosophy of nature, and I will return to it in the next chapter.

Nietzsche's turn to cosmology is largely due to Schopenhauer. As Parkes explains,

> An immediate prototype of the idea of a transpersonal or cosmic will is to be found in Schopenhauer, who argues in *The World as Will and Representation* that the entire world is basically will, as manifested in phenomena such as gravity, magnetism, and the lifeworld that drives plants, animals, and human beings. The human will is simply a more highly developed form of the basic force of the universe. Though Nietzsche's idea of the will to power is more complex, he follows Schopenhauer in understanding will cosmically and non-anthropocentrically.[51]

Parkes adverts to Schopenhauer's influence to ward off two common misinterpretations of the will to power: that "the 'will' of will to power is not the kind of willpower exerted by the human 'I' or ego; nor is the 'power' any kind of brute force exercised by human beings."[52] These are precisely Heidegger's charges, and as I discuss below, this is why he views Nietzsche as the heir of Descartes' philosophy.

But how is Nietzsche's view of will "more complex," as Parkes claims? The tendency of Schopenhauer's monistic view was to treat individuation not only as an illusion, but as the cause of suffering; the perspective of each being is a restriction, a prison in which it is alienated from its true nature. The best we can do is renunciation, to detach ourselves from the illusions that obscure the true unity of the world as will. Much like the Hindu philosophy that partly inspired it, Schopenhauer's view sees the world of form as an illusion or representation and the true world as unified. Nietzsche was unsatisfied with this view not only because of the ascetic morality it grounded (a morality that he saw as merely another mutation of the "ascetic ideal" and "bad conscience" that frustrate life) but because it did not do justice to individual differences and, despite is monistic surface, actually harbored a covert dualism between appearance and reality. As Walter Kaufmann provocatively suggests, "Nietzsche's will to power differs from Schopenhauer's will, much as Hegel's Absolute differs from his predecessors', Schelling in particular."[53] Just as Hegel strove for an ultimate principle that was not hiding "behind" but was literally incarnate in the manifest world, so Nietzsche rejects Schopenhauer's denigration of the "this-worldly." Put another way, Kaufmann explains that Nietzsche's ultimate principle "has an inherent capacity to give form to itself" and "in overcoming or sublimating itself, it appears in a strange dual capacity. It is both that which overcomes (e.g., reason) and that which is overcome (e.g., impulse). In Aristotelian terms, it is both matter and form; in Hegel's, it is both 'substance' and 'subject.' "[54] This means that, though Nietzsche's cosmology is monistic, it is a *developmental* monism: the will to power manifests itself in progressively more complex levels of form. This will be tremendously important when we treat his views of evolution and value in the next chapter, in large part because it highlights Nietzsche's reticence to postulate some mysterious force "behind" phenomena to explain them, for example, vitalism.

Nietzsche's motivation for undertaking a cosmology is to overcome nihilism. I think he would agree with Hans Jonas's framing of the problem of nihilism as reported by Lawrence Vogel: "[T]he metaphysical background of the nihilistic situation, according to Jonas, is the dualism between humanity and nature."[55] The strategy, then, is to side with positivism or scientific naturalism in rejecting traditional metaphysics, but to reject positivism's denial of soul or transcendence in nature, and to insist, with Aristotle, that something like soul pervades all living things. Here, we begin to see the resonance between Nietzsche's naturalism and the neo-Aristotelian naturalism Heidegger sketched earlier on.

The problem is that, in Heidegger's later interpretation, when Nietzsche is forced into the narrative of nihilism and the history of being, Heidegger interprets this cosmology as a kind of anthropocentric projection of human categories onto all beings. So Nietzsche is not, say, being more faithful to the phenomena by rejecting mechanistic materialism and imbuing natural beings with something like soul, but he is actually distorting them by projecting a historically contingent and thoroughly modern form of subjectivity, the will, onto them, and more importantly, he is overlooking the ontological difference between being and beings. Will to power, far from being the process of being's emergence, abiding, and withdrawal, is now interpreted as an all too human subjectivity trying to gain complete control of beings. It is the hallmark of humanism: denying a "vertical" transcendence of being opening up the world and doubling down on a "horizontal" transcendence of humanity progressively mastering nature through technology. Somehow, Nietzsche's attempt to overcome the nihilism implicit in the modern, mechanistic, materialist view of nature backfires and inadvertently greases the wheels for the final stage of nihilism, the *Gestell* in which even subject and object are dissolved in the self-enclosed system, the will to will that has no aim but to replicate itself. The overman, rather than freeing the last man from the cave, seals him more squarely within it. For Heidegger, Nietzsche's solution to nihilism—to create new, life-affirming values to replace the old, life-negating values—is, in Davis' perfect simile, like trying to "put out a fire with kerosene."[56]

Now let us look in more detail at Heidegger's analysis of will to power. Bret Davis has shown that this has to be seen in the context of Heidegger's own understanding of the will, which underwent a decisive turn in the period surrounding the Nietzsche lectures. In *Being and Time*, he writes, Heidegger "oscillates between embracing a resolute willing as the existentially decisive moment, and proposing that a shattering of the will is most proper to Dasein."[57] Hannah Arendt echoes the claims of Krell and Taminiaux I cited above when she charged that in the first Nietzsche lectures, the will takes the place of care in *Being and Time*. As Davis observes, "Heidegger largely 'goes along with' [Arendt's expression] Nietzsche's positive assessment of the will, going so far as to identify the will with his own key term 'resoluteness' from *Being and Time*, and with self-assertion, a key term of his Rectoral Address."[58] But in the second volume of the Nietzsche lectures, Heidegger claims that the will intrinsically constricts our access to beings and constrains their possibilities, reducing them to mere projections of our own designs, that this is the dominant understanding of being in the age of technology, that Nietzsche's philosophy of will to power is the philosophical foundation of this understanding, and that the rise of the will is the telos of the entire Western metaphysical tradition. Davis neatly summarizes the link between the will, nihilism, and the history of being: "Ultimately, in the age of nihilism as the most extreme 'epoch' of the history of being . . . beings are produced, ordered about, and distorted within an enframing set up and driven by the technological 'will to will.'"[59] Here again, we see the crucial distinction between "historical nihilism" and "ontological nihilism." The latter is common to all epochs, since it flows from the nature of being itself, while the former is the consequence of neglecting the latter (yet this neglect is somehow not a decision on the part of humans). So Heidegger's approach to the will to power is severely shaped by his critique of the will.

Heidegger's critique of Nietzsche's will to power centers on what Davis calls "ecstatic incorporation." The will to power, no matter the form in which it is found, is inherently self-enhancing and self-preserving and has no determining ground other than itself. All beings have certain conditions for their enhancement and growth, on the one hand, and certain conditions for their preservation and stasis, on the other. They are required to maintain the perspective or clearing they open up. These conditions do not exist "in themselves" prior to the emergence of that being; rather, they co-emerge with the coming into being of that being—they are thus contingent or conditioned conditions, but they *become* necessary once the being that depends on them has achieved a reliable degree of stability. These two drives—to enhance power and preserve it—require each other. The will is ecstatic insofar as it is by nature always already beyond itself, but this going beyond is only in order to secure itself; but it seeks to secure itself only in order to further enhance itself. This is why Nietzsche insists, "A living thing seeks above all to discharge its strength—life itself is will to power; self-preservation is only one of the indirect and most frequent results."[60] Keep in mind that Nietzsche is not talking about the will as a "faculty of the soul" but about life as such. It is not as if one can speak about an organism or being, on the one hand, and its willful activities and power, on the other; power ultimately refers to the unstable equilibrium of a being's bundle of drives in relation to themselves and to other bundles of drives, and to strive for a perfect equilibrium is to deny the very energies that make life possible. The "last man" is the being that lives only to last—he has no aim or purpose but to keep on living—and that is precisely what Heidegger will come to call the "will to will," the essence of historical nihilism. This process of expansion and consolidation aims at nothing beyond its own perpetuation. The opening out to the world is only for the purpose of enclosing more of the world within its sphere. As Davis explains, "Despite its ecstatic character, the will is after all, it would seem, a kind of 'encapsulation of the ego,' not, to be sure, in the aggressive sense of shutting out the world, but in the aggressive sense of expanding the territory of the ego to include the world in its field of power. . . . The will, in willing itself, reaches out to the world as something it 'posits.' "[61] Compare Heidegger in the following passage: "It is through human re-presenting that nature is brought before man. . . . Man places before himself the world as the whole of everything objective. . . . Man props up the world toward himself, and delivers nature over to himself."[62]

Now let us look at a rough sketch of Nietzsche's own account of nihilism and then shift to Heidegger's critique of the concept of value. The connection between will to power, values, and nihilism is roughly as follows. As noted above, all beings have certain conditions for their preservation and enhancement. These conditions are what Nietzsche calls values. So in this sense, all nonhumans posit values, since they are all essentially will to power, yet only humans can become conscious of their value-positing capacity. But since humans are the inheritors of the drives of prior forms of the will to power, their most basic and primitive projections of value are relatively unconscious, automatic, and naïve, and they thus assume that those values have an objective referent. Rather than see their value schemes and meanings as temporarily useful interpretations relevant and relative to a particular context—as creative responses to contingent life conditions—they take them to correspond to some sort of cosmic *logos*, such as the tao or nature. And Nietzsche's nearest definition of nihilism is that the highest values devaluate themselves, a process he deems the "Decline of Cosmological Values." It is crucial to point out, as Heidegger does, what Nietzsche intends by "cosmology" here: "Here cosmos does not mean 'nature' as distinct from man and God; rather, it signifies the 'world,' and 'world' is the name for beings as a whole. . . . [It] designates the widest circle that encloses everything that is and becomes. Outside it and beyond it nothing exists."[63] So when Nietzsche laments the decline of cosmological values, he is also referring to the collapse of a compelling cosmology. It is the sense—the "psychological state"—that the world lacks meaning, order, purpose, and unity. As Lampert puts it,

> If nihilism means that "the highest values devaluate themselves," then the highest values are themselves the cause of nihilism. Nihilism, Nietzsche maintains, is a "psychological state" of despondency reached when the highest values which "project" some value on the world are "pulled out" so that the world looks valueless. Nihilism as a sense of meaninglessness is a consequence of having believed in a meaning that is not there.[64]

So Nietzsche's claim that the realization that the world has no any meaning or value "in itself" is a calamity only against the background in which it is assumed to have one. It does not follow that there *is* no meaning or value, only that our prior estimations of it were largely a product of our own prejudices and were refracted through our anthropocentric prisms. Otherwise, Nietzsche's call for more life-affirming values and claim that the overman is the "meaning of the earth" are nonsensical. The problem is values that are grounded in what Heidegger calls "the supersensuous," in a fictional world beyond the natural one.

It is in this important sense that Nietzsche considered positivistic science an advance over premodern cultures mired in magical and mythical worldviews because it drains the world of spirits, gods, and "supernatural" forces that are, in truth, mere psychological projections. The strength of premodern societies was that they were bound by the spell of cosmological values, a taken-for-granted worldview that provided meaning and orientation for a people and drove them to sacrifice and strive for something beyond themselves. As Julian Young explains, "In the pre-modern era one had at one's disposal an ideal of virtue which enabled one to despise a life given over to mere contentment or—Nietzsche's general synonym—'happiness.'"[65] However, these cultures exist in a state of "latent" nihilism, since they have not yet come to terms with the fact that their values are human made. Their scope may be cosmic and permanent, but the conditions in response to which they were formed are local and transient. Positivism bursts this bubble, but the price of self-consciousness is the disorientation and despair that come from realizing that life has no completely readymade meaning or value in itself, that we must create it by our own lights and adopt a global perspective. Yet positivism has its own problems: it thinks it has skimmed off the mythological dross of less enlightened ages and broken through to the brute, value-free facts of reality, but Nietzsche argues that it is merely another world interpretation and that living *is* valuing. And the values promoted in the modern world—self-preservation, the search for scientific truth, the pursuit of "happiness," and the domination of the earth—are, Nietzsche insists, ultimately life negating. These produce a race of last men who no longer seek any great goals, but work only for their own preservation; one-half of the essence of life, self-enhancement or -transcendence, is effectively repressed. So Nietzsche's reading of modernity is a "decline" in the sense that it is a case of "arrested development." His vision of the future involves supplying the "1,001st goal," a vision that has the global scope of modernity (modernity being the current stage, which has no common goal) but that has the narrative power of premodernity.[66] This vision would dehumanize nature (strip it of the anthropomorphic projections that bedeviled premodern worldviews) but also renaturalize humanity (cease to conceive of it with supersensuous notions such as spirit or immaterial soul).

Here it is helpful to lay out the main distinctions in Nietzsche's account. There are three dyads: passive/active nihilism, incomplete/complete nihilism, and weak/strong nihilism. These are all basically synonymous, as Lampert explains: "For Nietzsche, the first half of each of these disjunctions indicates a nihilism that recognizes the devaluation of values but is unable to counteract it. The second half of each, on the other hand, is Nietzsche's own form of nihilism which he anticipates will be the nihilism of the future."[67] So we have roughly three stages here that have a somewhat Hegelian structure, as Lampert notes: "[T]he will to power expresses itself as *value bestowing*, as *value destroying*, and finally as *value bestowing in a new sense* based upon the recognition of will to power as the essence of things."[68]

The first stage comprises two parts. The first part is the initial establishment and faith in a set of values understood as eternal and anchored in the cosmos. This period covers the better part of human history. The second part is the dawning realization that these values cannot be maintained once humanity emerges out of its anthropomorphic childhood. This involves the inevitable collapse of those values and the futile attempt to sustain them; but this has negative consequences: disorientation, despair, pessimism, and bad faith.

The second and third stages fall under so-called active nihilism. The second stage is negative and critical and refers generally to a modern, scientific, positivistic program of "debunking" and "demythologizing" traditional interpretations of the world, in addition to the unmasking of wornout values. The active nihilist expedites the process of decline, but only to pave the way for the third stage, the establishment of more authentic values. The active nihilist recognizes that value-bestowing is unavoidable and that the positivist's attempt to be value neutral harbors its own value scheme. The active nihilist therefore attempts to value in a way that more accurately reflects nature as it is, rather than in light of an ideal world that distorts nature through an anthropocentric lens.

These three stages can be correlated with Nietzsche's metaphors of the camel, the lion, and the child from the section called the "On the Three Metamorphoses" in *Thus Spoke Zarathustra*. The camel is the beast of burden, the person that uncritically accepts the values of her people as eternally valid. The lion represents the emergence of freedom, individuality, and self-consciousness, the power to say "no" to conventions, laws, rules, and morality. This is a kind of negative freedom at odds with fate; it harbors resentment for the "it was" of the past, that before which it is impotent. The child, however, represents a creative spirit that is capable of generating its own values and doing so without resentment of the past or others or existence itself, that lives for the sheer joy of living and affirms the whole of reality. It throws itself completely into worldly projects with full knowledge of their futility. It has died to the dream of permanence. This figure embodies a kind of positive freedom at peace with fate and is encapsulated in Nietzsche's doctrine of *amor fati*. Aspects of existence that were previously rejected and airbrushed out of the "true world"—suffering, mortality, and temporality—are now embraced as gifts.

Now we need to look more pointedly at Heidegger's main problem with Nietzsche's account: the reliance on the concept of value. In the first chapter, I discussed Heidegger's aversion to value thinking, which was largely a reaction to the Neo-Kantian philosophy that ruled his own day. This aversion deeply colors his encounter with Nietzsche. As Krell observes,

> Heidegger's allergic reaction to Nietzschean "transvaluation of all values" derives partly . . . from his own rebellion as a student and young Dozent against the influential neo-Kantian *Wertphilosophie* of his mentor, Heinrich Rickert. . . . In valuative thought Heidegger sees the major obstacle to Nietzsche's advance beyond metaphysical modes of thought. The project of transvalution deflects and distracts Nietzsche from the questions of being, truth and the nothing.[69]

The mortal sin of Nietzsche's philosophy is that he reduces being to value, to that which is reckoned, calculated, and projected by humans. Heidegger often cites Nietzsche's definition of value from *The Will to Power*: "The viewpoint of 'value' is the viewpoint of conditions of preservation and enhancement with regard to complex constructs of relative life-duration within becoming."[70] By tying value to the notion of viewpoint, Heidegger reasons that evaluation is a process of fixing, reckoning, or calculating on the part of a subject, in the same way that he insists that time is reckoned and categorized in terms of presence in *Being and Time*. As Parvis Emad explains, "To Heidegger, Nietzsche's concept of value is the most extreme expression of constant presence, the mode of time to which Western metaphysical thinking has been continuously oriented."[71] This is strange, given that Nietzsche presented himself as a philosopher of becoming par excellence, rather than being. But Heidegger returns, again and again, to Nietzsche's insistence that the will to power strives to stamp becoming with the character of being *via* valuing. If there is one quote from the Nietzschean corpus that motivates Heidegger's interpretation, it is this: "[T]o stamp becoming with the character of being—that is the supreme will to power."[72] Heidegger cites this passage repeatedly throughout the Nietzsche lectures in order to buttress his thesis that despite Nietzsche's insistence that being is "a vapor and a fallacy," and no matter his attempts to overcome metaphysics, Nietzsche was unconsciously prisoner of the Greek interpretation of being as constant presence. This is why Heidegger claims that however much Nietzsche tries to escape from metaphysics, he reinscribes fundamental metaphysical concepts by reacting to them. For Heidegger, the categories of will and value are part of the problem of nihilism and cannot be part of the solution.

But Nietzsche thinks that the problem is not valuing itself, but *how* we valued up until now. Heidegger makes an important point:

What is untrue and untenable about the highest values hitherto does not lie in the values themselves, in their content, in the fact that in them meaning is sought, unity posited, and truth secured. Nietzsche sees what is untrue in the fact that these values have been mistakenly dispatched to a realm "existing in itself" . . . whereas they really have their origin and radius of validity solely in a certain kind of will to power.[73]

Put this way, Nietzsche's project of returning to the sensuous appears quite similar to Heidegger's phenomenological mission to avoid positing free-floating metaphysical concepts and returning to the things themselves. Yet Heidegger interprets Nietzsche's attempt to rehabilitate life and the sensuous as a reduction of the human essence to animality, of intentionality to "drives," and of being to beings.

III. Additional Problems with Heidegger's Interpretation

One of the strangest things about Heidegger's interpretation is that he conflates a *particular form* of the will to power with the *essence* of the will to power itself. At times, he alleges that Nietzsche's ideal for humanity, the overman, is a purely selfish being bent on the relentless acquisition of power and is the mindset that sets in motion and perpetuates the regime of modern technology. Heidegger's overman is the humanist par excellence, in Heidegger's pejorative sense of that term: having forgotten the source of his being, the humanist is enslaved to the delusion that he completely creates himself and that humanity is the source of its own significance. This being only sees others and nature as obstacles to and fodder for his own growth and preservation and uses his reason only to calculate the means to securing his own advantage.

This interpretation is strange because Nietzsche never tires of lambasting what he sees to be the psychological type most emblematic of modernity: a being that lives only to consume, that feeds off the base and baseless opinions of the populace, that prides itself on a sham individuality, that is ignorant of its history and the origins of its way of life, that believes humanity is the final stage of evolution and that modernity is the end of history, that subscribes to a vision of progress as the continual exploitation of nature to increase human comfort and alleviate pain and suffering, that aims only to "last," but not for any particular reason. As Dick White observes, "Clearly, what Heidegger is describing in his meditations on technology is the complete triumph of the slave/last man, which Nietzsche had already alluded to in his diagnoses of contemporary values and modern man."[74] And as I mentioned above, many of these characteristics also apply to Heidegger's portrait of inauthenticity, of Dasein as *das Man*, yet both Heidegger and Nietzsche maintained that humanity has other possibilities at its disposal. Indeed, Zimmerman points out that earlier on, Heidegger avers, "If will to power is understood rightly, not as capricious domination but rather as bringing-forth, then art can be understood as letting things be. Art [, Heidegger says] 'is an irruption by the man who knows and who goes forward in the midst of *physis* and upon its basis."[75] Here, will to power is "*physis* friendly." Moreover, at times Heidegger speaks highly of the overman. In *What Is Called Thinking*, he writes, "The thing that the superman discards is precisely our boundless, purely quantitative nonstop progress. The superman is poorer, simpler, tenderer and tougher, quieter and more self-sacrificing and slower of decision, and more economical of speech"[76] (*WIT*, 69). So it is strange that after the first Nietzsche lecture Heidegger reduces one manifestation of will to power to will to power itself. White summarizes his move: "[I]n the later essays on Nietzsche, the will to power is grasped in purely slavish terms, as a blind appropriation that aims to reduce everything to a uniform level so as to control it more effectively . . . In this way, the actual form of the will to power—technology—is treated as if it were the necessary result of value-positing as such."[77]

The problem, however, is that Heidegger seems to neglect the very different notions of power operative in Nietzsche's psychological types. Only the expression of power in the last man resembles the consumptive and acquisitive conception of power Heidegger descries in the *Gestell*; only this form of the will to power can be characterized by what Davis terms "ecstatic-incorporation." Nietzsche considers this an immature, unhealthy, weak, slavish, and arrested form of the will to power that measures the world by its own ego. As White explains, "the value-creation of the slave is inspired by vengeance, weakness, and fear; his values are primarily the values of utility . . . which serve to normalize society by suppressing individual difference. . . . This will to power is a will to possess power, a will to assimilate and control all that is outside of itself."[78] The mature form of the will to power has a completely different conception of the purpose of (and an entirely different motivation for) power-enhancement and preservation: it works to challenge and cultivate others to reach their highest potential. *That* is what guides the higher form of valuation, not the rapacious amassing of power for either the individual or for humanity as such. White puts his finger on what I take to be one of the central problems not just with Heidegger's take on Nietzsche, but with his own approach to nihilism: "Heidegger is led to reject the entire ethical/practical dimension of Nietzsche's philosophy . . . T]he rejection of the will to power *en bloc* effectively closes the door to every practical response to nihilism, as already engaged in the original sin of value-positing."[79] I would add, moreover, that it also closes the door to any environmental ethic. Despite Foltz's valiant attempt to retrieve an ethos of "dwelling" on the earth from Heidegger's work and Julian Young's insistence that Heidegger espouses some sort of "holy values" anchored in nature, these ideas seem incapable of generating any ideas more specific than "letting things be," which do not seem to provide the kind of discrimination called for in an environmental ethic. I thus concur wholeheartedly with White's conclusion:

> While Nietzsche can evaluate contemporary existence by making constant distinctions, geneaologies, and diagnoses of will to power and its various types, Heidegger's indiscriminate rejection of will and value bring him into an undiscriminating encounter with the modern world. . . . While Heidegger objects that Being has been "made into a value," his response of rejecting valuation in its entirety cannot be accepted.[80]

An underlying problem here is that both Heidegger and Young seem wedded to what we might call an anthropocentric reading of Nietzsche's theory of valuation. According to this view, values are purely human constructs "projected" onto a value-neutral, inherently meaningless universe. But Nietzsche explicitly criticizes this positivist view of nature as a modern prejudice and argues that it is itself nihilistic. More importantly, Nietzsche holds that, as even Young admits, "to live is to value," and he does not mean life in a merely human or existential sense. Indeed, life for Nietzsche is something like one of the medieval transcendentals; it is basically convertible with being. What this ultimately entails is a theory of value in which all beings value in some sense. In terms of their power enhancement, they posit values; in terms of their power preservation, they possess value. So, contrary to the popular view that Nietzsche reduces all value to human willing—and, to be fair, he sometimes speaks as if this were the case—there are parts of his philosophy that suggest that he offered a nonanthropocentric or perhaps cosmocentric theory of value. And the major point here is that the problem is not value positing or value hierarchies or willing as such—indeed, Nietzsche's whole point is that these are all unavoidable—but only *particular forms* of them. By attempting to avoid all of these things,

Heidegger makes himself vulnerable to pernicious forms of valuation and hierarchy, not to mention dubious master narratives of historical decline; by turning on Nietzsche, he turns away from those naturalistic possibilities in his own earlier thought.

Another feature of Heidegger's interpretation is his belief that Nietzsche was attached to a reified conception of the will inherited from the modern metaphysical tradition. As Krell notes, according to Heidegger, "will to power derives from Leibnizian *vis* and *appetitus*, and from the interpretations of will in Kant etc."[81] Heidegger makes the provocative claim that Nietzsche is unconsciously in league with Descartes: "Without being sufficiently aware of it, Nietzsche agrees with Descartes that Being means 'representedness,' a being established in thinking, and that truth means 'certitude.' "[82] Nietzsche rejects Descartes' interpretation of certitude as secured through the immediacy of an intellective act and claims that this is a function of a "will to truth," one of the forms of the will to power. Certitude is instead mediated by an act of will, which is to say that it is willed because it is valuable for the preservation of a certain form of life. Hence Heidegger claims that "Nietzsche refers the *ego cogito* back to an *ego volo* and interprets the *velle* as willing in the sense of will to power."[83] Nietzsche, he thinks, is somehow more Cartesian than Descartes because he projects a narrowly human mode of willing and knowing—the self-assertion of the "I think"—on to all beings, resulting in a panpsychism that is actually the most extreme form of anthropocentrism; there is no longer even a world of *res extensa* outside of the ego, only "our world of drives and passions." Heidegger is referring here to the quote from *Beyond Good and Evil* cited above, where Nietzsche posits that the same drives that compose human psychic life go "all the way down" even beneath the so-called organic level. This is the strategy Heidegger uses to claim that Nietzsche's will to power is not the counterpoint but the acceleration of Descartes' campaign to make humanity the "master and proprietor of nature."[84] This from a thinker who admonishes us to "Try taking away the phantasm and the entire human contribution. . . . If you could forget your heritage, your past, your training—your entire humanity and animality!"[85] Based on passages like this, Parkes wonders whether for Nietzsche "it may be after all possible to check that ancient positing, perhaps through some kind of phenomenological *epoche*, and let natural phenomena . . . simply show themselves, from themselves—perhaps even as they are in themselves?"[86]

Nietzsche's views of the body and consciousness do not resemble Cartesian concepts in the least. As Schrift points out, "In Nietzsche's writings, one finds an extended critique of the existence of the 'will.' For Nietzsche, the existence of a simple, solitary 'will' is a linguistic fiction which arises from our applying to diverse impulses a single name."[87] The human practical need to schematize and categorize leads us to mistakenly reify a complex of drives into a single, substantial entity. Rather than adopting an entirely new language to talk about human existence, as Heidegger does, Nietzsche attempts to describe the life of drives in psychological and biological terms. Every entity is a constellation of drives in tension with each other, and all drives value. The first part of *Beyond Good and Evil* is rife with the idea that there is no I-substance or -subject. Two representative passages:

> When I analyze the process that is expressed in the sentence, "I think," I find a whole series of daring assumptions that would be difficult, perhaps impossible, to prove; for example, that it is *I* who think, that there must necessarily be something that thinks, that thinking is an activity and operation on the part of a being who is thought of as a cause, that there is an ego, and . . . that I know what thinking is."[88]

> [A] thought comes when "it" wishes, and not when "I" wish, so that it is a falsification of the facts of the case to say that the subject "I" is the condition of the predicate "think." *It* thinks; but that this "it" is precisely the famous old ego is, to put it mildly, only a supposition, an assertion, and assuredly not an "immediate certainty." . . . Even the "it" contains an *interpretation* of the process, and does not belong to the process itself.[89]

What Heidegger really objects to, as we will see in chapter 7, is the characterization of life and beings in terms of "impulses" or "drives." He thinks this involves the "animalization" of man and, consequently, the refraction of being through this distorting prism so that both the dignity of humanity and the gift of being are obscured. As Zimmerman notes, for Heidegger, Nietzsche's "attempt to halt physiological degeneration [i.e., nihilism] by 're-animalizing' man proved to be the final stage in the ontological degeneration of man."[90] Heidegger reduces the overman to the "underman," an inauthentic comportment, an animal captive to its own drives.

One final point about Heidegger's interpretation of the will to power is that he claims that, despite Nietzsche's utterances to the contrary, the connection between the will to power and the eternal recurrence of the same—namely, the drive to stamp becoming with the character of being, or permanence—shows that Nietzsche was still imprisoned by the thought of being as presence, that is, metaphysics. But the reason Nietzsche dismisses being as "a vapor and a fallacy" is that he agrees with Heidegger about the inadequacy of substance metaphysics, yet he does not see that as grounds for completely abandoning the attempt to give an account of the ontic order. Nietzsche's will to power is intended as a metaphysical principle that characterizes all beings—including humans—whereas Heidegger, preoccupied with the hermeneutic issue of access to beings, is reticent about the possibility of any such metaphysics. Nietzsche does not think that the different "regions" of being can be so easily separated; as we will see in the next chapter, this is why he equated being with life and sees the latter as forming an ontological continuum that integrated the psychological, biological, and inorganic orders. Another way of saying this is that Nietzsche attempts a metaphysics of nature, an *alternative* metaphysics, whereas Heidegger believes that any such attempt is necessarily anthropocentric. Zimmerman summarizes Heidegger's position in this respect:

> Heidegger concluded that the intrinsic possibilities of living things are discovered, not created, in the historical world. If living things did not somehow contain their own *intrinsic limit and measure*, there would be no basis for objecting to the technological disclosure of things. . . . Heidegger's difficulty, however, was how to speak of the intrinsic measure and limit of living things ('the hidden law of the earth') without resorting to one of the foundationalist doctrines of productionist metaphysics. . . . *Unwilling to appeal to either metaphysics or to science*, Heidegger concluded that the 'hidden law of the earth' is an impenetrable mystery."[91]

Conclusion

What remains is to determine what in Heidegger's thinking led him down this path. I contend that this is his conception of life. In the next chapter, I contrast this with the alternatives of Nietzsche and argue that his conception of life, intentionally formulated in response to the problem of nihilism, points to a more viable view of nature that is actually congruent with Heidegger's earlier view. Whereas Nietzsche elaborates a robust philosophical biology, engages evolutionary theory, and formulates a value theory rooted in the natural world, Heidegger abandons these paths. But in doing so, he turns his back on the nascent naturalism sketched in his earlier work. Nietzsche's conception of life can help us get a sense for where Heidegger's thought might have led had he pursued these earlier lines of thought and can supplement Heidegger's rich investigations.

7

Naturalizing Nietzsche

Life, Evolution, and Value

In this chapter, I draw on ideas from Nietzsche's thought to develop the nonreductive naturalism sketched in Heidegger's earlier work on Aristotle and Uexküll and pursued by later phenomenological thinkers such as Hans Jonas and Evan Thompson. I focus on Nietzsche's view of living being, his incorporation of evolution, and his value theory. This kind of naturalism is "life affirming" in the sense that it recognizes life as an autonomous kind of being irreducible to physiochemical properties and mechanistic causality. It holds that humanity is continuous with animal life and subject to evolutionary forces yet resists the mechanistic, materialist interpretation of evolution. And finally, it rejects the value-free vision of nature found in modern science, holding that all living things value in some sense.

In the first section, I lay out some of the basic themes at play in interpreting Nietzsche naturalistically. There are a variety of views on whether and to what extent Nietzsche is a naturalist, so I begin by surveying the major positions and controversies.

In the second section, I lay out Nietzsche's naturalism by focusing on his view of biology. We have to begin with the biology because it is Nietzsche's point of entry for anchoring value in the natural world; it is by liberating biology from mechanism and recovering a nonmetaphysical form of teleology that we will attain an account of nature as value laden. In this section, I draw on several thinkers to help clarify and elaborate the nonreductive naturalism I have attempted to cull from Nietzsche and Heidegger. Hans Jonas offers an "existentialist interpretation of biological facts," a phenomenological naturalism that underwrote an ethic anchored in the natural world. In a similar fashion, Evan Thompson has recently put forth an ambitious theory of "mind in life," a synthesis of biology and phenomenology. Both thinkers argue stridently against materialism and offer an alternative naturalism that entails a conception of natural value. I also draw liberally from John Richardson's reading of Nietzsche's biology not only because of its clarity and rigor, but also because of its rarity. Despite the centrality of biology to Nietzsche's thought, there are surprisingly few in-depth analyses of it; indeed, even a recent anthology devoted to Nietzsche's naturalism scarcely mentions his views on biology and evolution. My main purpose in the following section is to clarify Nietzsche's specific form of naturalism and his positions on drives, values, evolution, teleology, and selection. Once we have this positive vision of nature in place, we can turn to its implications for environmental ethics.

I. Nietzsche's Naturalism

For Nietzsche, the rise of positivism—which he regarded as the final stage in the "history of an error" and the culmination of metaphysics—meant the "decline of cosmological values." The stone of metaphysics cast upward in Platonism, Christianity, and idealism, which projected a supersensuous, supernatural world as the standard by which this world was to be judged and served as the locus of eternal values, had to come down in modern science and positivism, which declared that mere nature—a closed, materialistic, meaningless, mechanistic order—was all there is. Nietzsche's pronouncement that God is dead meant that the supernatural order in which human beings placed their highest hopes and values never existed in the first place and that the dawning realization of this truth, when placed before the background of a nature without purpose or value, would lead to great confusion and disorientation about the meaning of human life.

But Nietzsche was no romantic. He believed that the neat cosmological orders of the past were also nihilistic, since they mistook local and contingent valuations for cosmic and necessary values and that the modern understanding of nature appears meaning- and value-less only against the background of mythology. He thus envisioned the possibility that after the painful process of critique and the overcoming of premodern prejudices, of bracketing and unlearning the unfounded positings of magic, myth, metaphysics, and even science, the way would be clear for a revaluation, one that was more attuned to nature as it is, not as the photographic negative of a projected ideal world. And given his conviction that life inherently values, that is has conditions for its own preservation and enhancement, his positive vision of nature includes a conception of natural value. Nietzsche's screeds against the "human, all too human" character of valuation are aimed at previous valuations, not valuation as such. So Nietzsche's genealogical unmasking of metaphysical values, while proximally intended to debunk cosmic or objective values as human projections, is actually ultimately geared toward arriving at a positive vision of nature as value-laden.

Nietzsche's task was thus to accept and digest the "dangerous ideas" of modern science—especially the theory of evolution—without succumbing to their nihilistic implications. As R. J. Hollingdale summarizes, "The sense that the meaning of the universe had evaporated was what seemed to escape those who welcomed Darwin as a benefactor of mankind. Nietzsche considered that evolution presented a correct picture of the world, but that it was a disastrous picture. His philosophy was an attempt to produce a new world-picture which took Darwin into account but was not nullified by it."[1] That, in a nutshell, is why the solution to the problem of nihilism involves the search for a new vision of nature, and why the pivotal concept is that of life. This is by no means an issue that has been settled by the neo-Darwinian synthesis. In 2003, in an article entitled "Darwin's Nihilistic Idea: Evolution and the Meaninglessness of Life," Tamler Sommers and Alex Rosenberg zero in on the connection between values and biology:

> Darwinism puts the capstone on a process which since Newton's time has driven teleology to the explanatory sidelines. In short it has made the Darwinians into metaphysical nihilists denying that there is any meaning or purpose to the universe, its contents and its cosmic history. But in making Darwinians into metaphysical nihilists, the solvent algorithm should have made them into ethical nihilists too. For intrinsic values and obligations make sense only against the background of purposes, goals, and ends which are not merely instrumental.[2]

Nietzsche took humans' natural and unavoidable capacity for valuation as a sign that valuation is intrinsic to life. Nietzsche attempted to "dehumanize nature" while "re-animalizing man," but without lapsing into scientific naturalism.

His own naturalism is hard to place and continues to be a matter of dispute.[3] It has been termed both a "naturalistic transcendentalism" (Ralph Acampora) and a "transcendental naturalism" (Keith Ansell Pearson). Graham Parkes sees Nietzsche as a panpsychist.[4] On the other end of the spectrum, Brian Leiter sees Nietzsche as more of a scientific naturalist.[5] Richard Schacht offers perhaps the most comprehensive definition of Nietzsche's naturalism. In Schacht's view, the "guiding idea" of Nietzsche's naturalism is this: "that everything that goes on and comes to be in this world is the outcome of developments occurring within it that are owing entirely to its internal dynamics and the contingencies to which they give rise, and come about (as it were) from the bottom up, through the elaboration or relationally-precipitated transformation of what was already going on and had already come to be."[6] Nietzsche's naturalism is thus intended to chart a middle way between a dualistic, transcendent metaphysics and a reductive, scientific naturalism. Schacht claims it is "minimalist" because it is only committed to the "guiding idea" mentioned above; it is "extended" because not only does it translates "man back into nature," but it sees humanity as transcending and transforming its animal inheritance; and it is "robust," "emergentist," and "historical" because, unlike scientific naturalism, it acknowledges the psychological, social, cultural, and artistic aspects of human experience, not just the physiological and biological aspects.

Schacht—correctly, I think—rejects what he sees to be Brian Leiter's scientistic interpretation, which holds that Nietzsche was committed to a "substantive" naturalism, a "scientific picture of how things work."[7] On the contrary, in Schacht's view: Nietzsche's task was to debunk not just the ontological pretensions of religion, but those of modern science as well. Though he does not mention it explicitly, Schacht zeroes in here on a key to understanding Nietzsche's approach to the interlocking themes of naturalism and normativity: nihilism. As Schacht notes, at *Gay Science* 373 Nietzsche clearly rejects a "'scientific' world interpretation" because it cannot account for things that are "*meaning-constituted*": "Nietzsche here has 'mechanistic' thinking specifically in mind; but his basic point applies to natural-scientific thinking more generally: such thinking is inherently *meaning-blind*."[8]

A curious, and in my view glaring, omission in Schacht's account—and in the volume devoted to Nietzsche's naturalism of which it is a part—is any substantive discussion of philosophical biology. The book's index contains no entries on biology or Darwin. Careful work has been done by scholars such as Gregory Moore and John Richardson on the importance, indeed centrality, of biological and evolutionary considerations in Nietzsche's work, especially his naturalism but including his value-theory. Daniel Dennett, for one, ultimately rejects Nietzsche's philosophy because it is incompatible with neo-Darwinian biology. Nietzsche, he thinks, is a pioneering philosopher precisely because he understood the philosophical implications of evolutionary theory; however, it is because Nietzsche (definitely) rejected Darwinian mechanism and (allegedly) embraced a form of vitalism that Dennett rejects his philosophy of the will to power. Even Richardson, who does his best to make Nietzsche's thought compatible with, or palatable to, scientific naturalism, concedes that Nietzsche's dominant tendency is to see the will to power, that is, drives, in Lamarckian terms, operating above and beyond natural selection *a la* the *Bildungstreib* of the romantic biologists of the nineteenth century. One of the questions I explore in the next section is whether or not Nietzsche is a vitalist.

He rejects the mechanistic view of the animal (and the mechanistic view of the inorganic world) advanced in modern science and attributes some degree of subjectivity, interiority, or self-organizing capacity to all living things. Zimmerman neatly frames his position:

> Nietzsche agreed with naturalism that otherworldly religious categories are not explanatory, but he also showed contempt for mechanistic naturalism, because it ignores purposiveness and will, central instances of the interiority that Nietzsche ascribed to everything in a way that seems consistent with a variety of panpsychism. Retaining a place for consciousness, soul, and spirit in his naturalism, Nietzsche ascribed to plants and animals an interiority that is overlooked in principle by the sciences that focus solely on mechanical behavior.[9]

Framing humans as animals is "reductive" only if one has previously reduced animals to extended matter governed by mechanical processes and divested them of any cognitive or affective capacities. So I suggest that Nietzsche's project falls within the category of nonreductive naturalisms, which Ted Benton defines thus:

> A non-reductionist naturalism, making use of the ideas of a hierarchy of more or less autonomous levels of organization of matter, each with its own, qualitatively new, "emergent" powers or properties has been one fruitful way of maintaining the insights of a naturalistic approach, without falling foul of what is valid in the anti-naturalistic critique. Such hierarchical, "emergent powers" ontologies enable their advocates to recognize in the various subject matters of the different natural and social sciences more or less discrete and autonomous object-domains, while at the same time making no concessions to spiritualistic, vitalist, or supernatural beliefs.[10]

What Nietzsche is doing, I propose, is trying to reconstruct the great chain of being without speculative supports in a way that is consistent with biological science. In so doing, he is trying to thread the needle that Heidegger identified in his own forays into biology: how to avoid the extremes of mechanism and vitalism.

II. Nietzsche's Philosophy of Biology

I submit that the key to Nietzsche's naturalism is his philosophy of biology, and the complexity of Nietzsche's biology is reflected in his ambivalence toward Darwin. There is no doubt that Nietzsche intentionally opposes himself to Darwin; much like his dramatic portrayal of himself as the anti-Christ or anti-Christian, he commonly labels his views as "Anti-Darwin."[11] His chief objections to Darwinism are that it prioritizes the species over the individual, that it has a one-sided emphasis on self-preservation, and that it mistakenly posits that selection favors the strong rather than the weak.

Nietzsche explicitly names Darwinism as a form of nihilism. His notion of the "last men" is no doubt deeply tied to the idea that modern humanity is undergoing a period of degeneration, cut off from its sources of vitality. As Gregory Moore has documented, "Spencer's 'ideal moral man' is the prototype for Nietzsche's last man. It has to do with Spencer's claim that evolutionary development aims at the prolongation of life."[12] Indeed, in section 373 of the *Gay Science*, Nietzsche lambastes Spencer's hope for an eventual "reconciliation of 'egoism and altruism,'" insisting that "a human race that adopted such Spencerian perspectives as its ultimate perspectives would seem to us worthy of contempt, of annihilation!"[13] Later in the same section, Nietzsche avers that "an essentially mechanical world would be an essentially meaningless world."[14] In this regard, he would seem opposed to evolutionary science, since it is life denying and seems to rob humans of meaningful goals.

However, as Moore points out, Nietzsche had a "lifelong fascination" with "the far-reaching implications of the modern evolutionary worldview for the traditional areas of philosophical inquiry. Indeed, the central project of his later thought—the much-vaunted transvaluation of values—rests precisely upon an appeal to the explanatory power of a newly confident biology."[15] He continues: "There can be no question that Nietzsche adopts a broadly evolutionary perspective: he believes in the mutability of organic forms; he sees morality, art, and consciousness not as uniquely human endowments with their origin in a transcendental realm, but as products of the evolutionary process itself."[16] But what sort of evolutionary view does Nietzsche embrace? It is clear that he rejects a mechanistic account in which the environment does all the "work" of selecting the traits and behaviors of the organism. And a materialism is out the question: Nietzsche regards mere "inert matter" not only as an abstraction from our experience of ourselves and of the world of living things, but as an inadequate explanation for the so-called inorganic world.

What about a teleological account? Nietzsche never tires of arguing that there are no purposes in nature, no natural kinds with fixed *teloi*, whether determined by God or nature; final causes are imputed by humans in order to make sense of the world.

Perhaps a version of vitalism? But precisely what is vitalism? According to William Bechtel and Robert Richardson,

> Vitalists hold that living organisms are fundamentally different from non-living entities because they contain some non-physical element or are governed by different principles than are inanimate things. In its simplest form, vitalism holds that living entities contain some fluid, or a distinctive "spirit." In more sophisticated forms, the vital spirit becomes a substance infusing bodies and giving life to them; or vitalism becomes the view that there is a distinctive organization among living things.[17]

Nietzsche's belief in a common life force that governs the growth and development of all living things would seem to place him in this camp. He read and drew deeply from a number of influential biologists of his day that would later be deemed, and dismissed as, vitalists. These theorists' views represented the anthropomorphism and lack of philosophical sophistication of which Heidegger accused Uexküll. Moore, for one, claims that "Nietzsche reiterates the many errors and misunderstandings perpetrated by his contemporaries. Like them, he dresses up metaphysical and anthropomorphic views of nature in the language of modern evolutionary biology. The will to power is essentially a *Bildungstreib*, as it were an amalgam of a number of competing non-Darwinian theories."[18] And Dennett, otherwise impressed with Nietzsche's appreciation for the power of evolutionary thinking, laments Nietzsche's resort to "skyhook hunger" by rejecting mechanism for the will to power.[19] Bechtel and Richardson relay vitalism's status in relation to mainstream biology:

> Vitalism now has no credibility. This is sometimes credited to the view that vitalism posits an unknowable factor in explaining life; and further, vitalism is often viewed as unfalsifiable, and therefore

a pernicious metaphysical doctrine. Ernst Mayr, for example, says that vitalism "virtually leaves the realm of science by falling back on an unknown and presumably unknowable factor. C. G. Hempel, by contrast, insists that the fault with vitalism is not that it posits entities which cannot be observed, but that such explanations "render all statements about entelechies inaccessible to empirical test and thus devoid of empirical meaning" because no methods of test, however indirect, are provided. The central problem is that vitalism offers no definite predictions.

However, they also point out that neither of these charges is valid, since many vitalists attempted to confirm their theories through experimental testing and were often motivated to do so because alternative mechanistic explanations were found to be empirically wanting. In other words, though mechanistic explanation was eventually proved superior through experimental means, the contrast between hard-nosed, empirical, experimental mechanists and speculative, unscientific, metaphysical vitalists is overdrawn.

The vitalist interpretation is not without merit. Moore's charge has teeth: he posits that Nietzsche was caught up in an emerging scientific paradigm that was later superseded. And indeed, insofar as Nietzsche embraces vitalism, his views ought to be rejected. However, given Nietzsche's resistance to positing abstract entities "behind" phenomena in order to explain them and his acute sensitivity for the anthropomorphic tendencies of philosophers, perhaps we should not be too hasty in branding him a vitalist. Moreover, I think we should suspend the orthodox assumption that anything that deviates from neo-Darwinism—mechanism, materialism, and scientific naturalism—is automatically spooky metaphysics unworthy of attention. There is certainly a serious tension in Nietzsche's thought between a more restrained, more scientific naturalism and a more ambitious, speculative view of nature. I submit that at his best, Nietzsche either is not a vitalist—in the pejorative sense of the term—or he is a vitalist in the minimal and restricted sense noted in Bechtel's and Richardson's definition I cited above: that organisms are "governed by different principles than are inanimate things" and that there is "distinctive organization among living things." In order to flesh this out, I draw below on John Richardson's meticulous reconstruction of Nietzsche's view of evolution and, later, Jonas's and Thompson's syntheses.

Richardson has dispelled much of the confusion over Nietzsche's account of evolution. I will reconstruct and supplement his analysis here because it shows how Nietzsche can be read as a nonreductive naturalist and how values figure into life. The key to Nietzsche's theory of evolution, Richardson contends, is his conception of drives. "Drives," Richardson writes, "are his principal explanatory tokens. He attributes drives to all life, and analyzes organisms (and persons) as complexes of drives."[20] Moreover, Richardson claims that the Darwinian dimension of Nietzsche's thinking actually renders it more plausible. He is getting at a difficult tension in Nietzsche's thought between a more metaphysical view of the will to power that falls prey to anthropomorphism or "power ontology" (Heidegger's and Dennett's view) and a more naturalistic view or "power biology" (Richardson's own position). Richardson thinks Nietzsche oscillates between these two views, but that the former is dominant; he is careful, however, to note that "the view is 'metaphysical' not in treating [will to power] as transcendent, or as knowable *a priori*, but as primitive—uncaused and unexplainable."[21] He is interested in the extent to which Nietzsche's philosophical biology can be naturalized, or rendered palatable for contemporary biologists:

> Nietzsche's degree of success in naturalizing his biology (and psychology, as likewise insisting on our willing power) depends on how far he can see this will to power as a selected product, not an ultimate principle or life force. However it's important to bear in mind that even when he does think of will to power as a prior such force, he still thinks of selection as shaping it. He still explains much or most of the character of organisms' and persons' drives by ways that selection has culled that *ur*-force into behaviors that can survive and continue. . . . He has naturalistic senses for the ways drives 'value' and are 'toward ends,' to precisely the extent that he explains them by selection.[22]

Richardson's project is to show that Nietzsche's metaphysical vision of the will to power still rests on a naturalistic foundation, even though it goes beyond it. While I agree with Richardson that we want to reject a sloppy panpsychism that unwarrantedly attributes mentality to nonhumans, I think he is too acquiescent before the hardcore naturalism of neo-Darwinism; as we'll see below, there are good reasons for resisting it. And Richardson acknowledges that Nietzsche's views on the social and psychological sources of valuation cannot be (completely) accounted for in Darwinian terms.

Before spelling all this out, though, let us take a closer look at Nietzsche's biology from five angles: his views on Darwin, teleology, drives, values, and the different kinds of selection.

1. Darwin

To begin, as we saw Moore point out above, Nietzsche embraces Darwin's basic idea that humans are the product of a natural process of terrestrial evolution, he agrees that natural selection plays a major part in determining organic forms, and he believes that much of human morality, religion, and culture can be understood in terms of this natural process. Indeed, Nietzsche was principally interested in doing what Darwin did only later in his career, in the *Descent of Man*: drawing the consequences of the evolutionary idea for human beings. It is tightly bound to his central motif of the death of God. As Richardson explains,

> Nietzsche associates with Darwin certain "critical"—skeptical and nihilistic—lessons. . . . He takes Darwin to have these critical consequences by his decisive step in naturalizing life—i.e., in explaining it by processes that are nondivine and indeed noncognitive. . . . Part of Darwin's insight is just evolution itself: species become, are created and destroyed, including the human species. But more important is his account of what drives that evolution: a struggle or competition in which all organisms—ourselves included—are engaged.[23]

So Nietzsche regards Darwin's discovery as a solid support for his general view of modernity and project of debunking false worldviews and values. Darwin's dangerous idea has to be digested.

As we turn to the disagreements, it is important to keep in mind that, as is sometimes the case with Nietzsche,[24] his knowledge of his subject—in this case Darwin—is gained second-hand: his main sources are Spencer and a number of Darwin's critics. It is therefore unsurprising that, as Moore notes, Nietzsche's view of evolution cleaves closely to Spencer's: "Like the activity associated with Spencerian evolution, the will to power is a development from the simple to the complex, and takes place . . . on a cosmic scale. Nietzsche's concept of *Entwicklung* [development] thus has more in common with Spencer's understanding of evolution than it does with Darwin's."[25] However, Richardson has carefully shown that a number of Nietzsche's disagreements with Darwin are baseless and that their views are actually in sync.

One major difference is that Nietzsche believes that Darwinism smuggles a moral prejudice into its understanding of life: a conception of progress in which organic forms develop toward increasing perfection, with humans as the crowning achievement of nature. This, he thinks, merely reflects the decadent and leveling spirit of modern culture, not the "things themselves." The spread of altruism, which Nietzsche regards as the latest incarnation of slave morality, results in the stifling of struggle and self-overcoming of distinctive and powerful individuals; it is not what it is presented as—the overcoming of egoism—but exactly its opposite—the justification of mass egoism. The well-intentioned attempt to stamp out brutish egoism actually conceals a subtler egoism, a tyranny against the instincts. The brutish animal spirits must be tamed by the hive-mind of the democratic, civilized, egalitarian order. Nietzsche thinks that this view of evolution, rather than connecting humans with their vitality and animality, actually represses them. So Nietzsche thinks that this view is nihilistic because it imputes a moral teleology to human history and then inflates this teleology to encompass the development of life itself; it is thus supremely anthropomorphic.

Another difference is that Nietzsche faults Darwin for framing the struggle of life as a struggle for *existence* in which the *physically* fittest specimens win out. As Richardson points out, Nietzsche "misreads Darwinian 'struggle' as physical combat, and 'fitness' as muscular strength. So he takes the latter to exclude all the indirect devices he labels 'cunning.' But of course Darwin makes clear that organisms struggle in many different ways; see, e.g., his account of the cuckoo's instinct to lay its eggs in other birds' nests."[26] Again, Nietzsche's misreading is probably due to his reading of Spencer, who famously coined the phrase "survival of the fittest." Nietzsche objects to "fitted-ness" because he thinks it is contaminated by the same moral prejudice mentioned above: the instinct to conformity, that the success of the organism lies in conforming or adapting itself to its environment, rather than creatively responding to and shaping it.

This brings us to a third disagreement, which has two related facets. Whereas Darwinism held that the stronger individuals succeed, Nietzsche thought the reverse: that over time, it is the weak that come to dominate. The strength in numbers of the herd retards the development of higher types. A corollary of this is that Darwinism conceives of evolution in terms of the preservation of the species, whereas Nietzsche sees it as geared toward the production of exceptional individuals. As he writes: "Fundamental error of biologists hitherto: it is not a matter of the species, but of bringing about stronger individuals."[27] At *WP* 685, he says that "growth in the power of a species is perhaps guaranteed less by a preponderance of its children of fortune, of strong members, than by a preponderance of average and lower types."[28] Indeed, for Nietzsche, evolution does not take place merely by dint of organisms reproducing and passing on their type with modifications.

There are two poles of evolution, the group and the individual, and only the latter truly evolves. Groups reproduce more and are more stable. But individuals demonstrate the inner dynamism, the struggle that leads to the creation of new, higher forms and the subordination of lower, older ones. As Moore explains, Nietzsche was strongly influenced by Wilhelm Roux, a student of Haeckel, who thought the internal struggles of organisms were the main cause of their form:

> Self-regulation is the mechanism by which the random variations produced by overcompensation are ordered and selected by the functional requirements of the whole. . . . The development of "aristocratic" hierarchies, in which the strongest parts of an organism subdue the weaker ones . . . with a more complex organic structure emerging through the subsumption of lower forms by higher ones; cells by tissues, tissues by organs, and so on.[29]

For Nietzsche, the strength of an organism consists in its ability to develop autonomously, not merely in reaction to its species or "society," as Moore notes: "The hallmark of an evolving, higher organism is its ability to regulate the internal relationships of its drives, now severed from a collective, superordinate identity."[30] The "herd" exerts a tremendous selection pressure that, though initially a creative transcendence of another, older "herd mentality," has outlived its usefulness, no longer fosters growth, and retards future development. The main point here is that newer, more complex forms are rare, more fragile, and less likely to be replicated.

I want to pause the discussion of Nietzsche's disagreements with Darwin here in order to start to tease together the positive alternative that his criticisms imply. What we see coming into focus in Nietzsche's view of evolution is a dialectical process taking place at all levels of organization: first, a creative interpretation that organizes the world in such a way as to foster the growth and preservation of the organism; second, this settles into a stable pattern or form of life that guides the development of subsequent organisms; third, life conditions change, and the pattern ceases to foster growth and becomes an end in itself, bent only on preservation; fourth, a new pattern more attuned to the present life conditions supplants it, and the process starts over again. For Nietzsche, "growth in life" means "an ever more thrifty and more far-seeing economy, which achieves more and more with less and less force."[31] If we can read "pattern" and "economy" here as more or less synonymous with "interpretation," "perspective," and "set of values," we begin to see how Nietzsche attempts to integrate biology, psychology, and values in a nonreductive view of evolution. At *WP* 636, he describes perspectivism: "My idea is that every specific body strives to become master over all space and to extend its force . . . and to thrust back all that resists its extension. But it continually encounters similar efforts on the part of other bodies and ends by coming to an arrangement ('union') with those of them that are sufficiently related to it: thus they then conspire together for power. And the process goes on."[32] And power here is the ability to delimit and inhabit a horizon, a creative capacity of life. This is what Nietzsche means by interpretation:

> The will to power interprets—it is a question of interpretation when an organ is constructed: it defines limits, determines degrees, variations of power. Mere variations of power could not feel themselves to be such: there must be present something that wants to grow and interprets the value of whatever else wants to grow. . . . In fact, interpretation is itself a means of becoming master of something. (The organic process constantly presupposes interpretations.)[33]

Now back to our discussion of the disagreements with Darwin. The fourth point of disagreement is that Nietzsche appears to have adhered to a version of Lamarckism, or the "inheritance of acquired characteristics." While Darwinism holds that organic forms are gradually built up over long stretches of time by the selective pressures of the external environment, Lamarck believed that traits could be modified through behavior and habituation within the lifetime of the organism and that such traits could be passed on to offspring. Nietzsche was attracted to this latter, more horizontal form of evolution because it was more attentive to the life, behavior, and development of the individual organism, instead of Darwinism's focus on the species and subjection of the individual to mechanical forces; on this point, his views track with Uexküll's.

The final and most important disagreement concerns two points: the instinct for self-preservation and teleology. Richardson points out that this disagreement has to do "with Darwin's stress (Nietzsche thinks) on survival or preservation, instead of on power or growth," and that "[Nietzsche] conceives [power and survival] to be competing answers to the question of the end or goal of life: he takes Darwin to be claiming that organisms are 'toward' survival, and he argues that organisms are directed toward power.

More specifically, he supposes that both of these are meant as goals of a 'will' or 'basic drive' of life, which is *zu* or *auf* or *um* them."[34] Nietzsche's position on teleology is difficult to pin down. At *Beyond Good and Evil* 13, he says

> Physiologists should think before putting down the instinct of self-preservation as the cardinal instinct of an organic being. A living thing seeks above all to discharge its strength—life itself is will to power; self-preservation is only one of the indirect and most frequent results. In short, here as everywhere else, let us beware of *superfluous* teleological principles.[35]

I think we should heed the word "superfluous" in this quote: Nietzsche wants us to be on guard against projecting unfounded goals or ends onto phenomena; he is not saying that teleological explanation can be done away with altogether. Or does he? Many passages—for example, "We have invented the concept 'purpose': in reality purpose is absent"[36]—emphatically deny purposes in nature. Nietzsche unquestionably rejects the classical model of teleology, which rests on a substance/accident and form/matter model of explanation, or a theistic account of teleology, in which the thing's form and end are patterned according to an idea in the mind of God, both of which posit a *fixed* end that guides a thing's development and behavior; indeed, it is this notion of natural kinds or essences that he takes Darwin's theory to have demolished. And yet, as Richardson wonders, "these rejections [of teleology as such] seem at odds with his insistence on a will 'to' power. What can that towardness be, if *not* an end-directedness?"[37] What is going on here? My view is that Nietzsche's rhetoric about teleology is, as on many issues, hyperbolic and that despite his critiques of previous forms of teleology, he does, as Richardson persuasively argues, embrace a qualified, more naturalistic form of the concept. As I discuss below, some form of teleology is demanded by Nietzsche's views on drives, values, and selection.

The second problem has to do with the status of the "goal" of self-preservation, and it connects to the major issue I mentioned above: whether and to what extent Nietzsche embraces a kind of panpsychism that illicitly imputes mentality to all living things. Richardson lays out the problem:

> Nietzsche seems to misread Darwinian survival as an "end" in a too literal sense: as the aim of a will or drive or instinct. . . . Nietzsche's terms "will" and "drive" suggest an intentional end-directedness—that either power or survival is an intended goal. And this in turn suggests that there is some kind

of representation of the goal, which picks it out in advance and steers behavior toward it. But Darwin's core point about natural selection posits no such "self-preservation drive," nothing that "aims" or "steers" organisms at reproduction. It rather describes a long-term structural property of evolution: traits that improve fitness tend to persist and accumulate; this mechanism, operating over long periods, explains organisms' most striking features. Those features have been designed for certain functions. . . . Those functions, and reproduction, are not represented goals, but the outcomes for which those biological features were selected. So it appears that Nietzsche offers power to replace survival in a role the latter was never meant to play.[38]

Nietzsche was likely led down this path of thought because of his concern that Darwinists were representing the evolution of life in terms of modern historical progressivism—such as Spencer's conception of the survival of the fittest, the idea that "later" is "better"—thus mistaking one of the effects of human history for the cause of the evolution of life. He warns us not to "set up terminal forms of evolution (e.g., spirit) as another 'in itself' behind evolution!"[39] Nietzsche thinks that though Darwinism is presented as mechanistic, it smuggles in a form of teleology—the will to life/existence/preservation—that he thinks is a degenerate, life-negating attitude, one that is actually divorced from the "drive-life" of living things. Moreover, the mechanist plays a shell game with value, meaning, and purpose. He takes them all away from nature, but then has to explain how they emerge for consciousness in mechanistic terms; and he cannot account for his own ability to give a meaningful account.

But this presupposes an alternative understanding of life's directedness. And the danger is that, as we saw Heidegger suggest in previous chapters, this alternative at times smacks of an anthropomorphizing panpsychism or vitalism. Indeed, at *WP* 636, we find: "[Physicists] left something out of the constellation without knowing it: precisely this necessary perspectivism by virtue of which every center of force—and *not only man*—construes all the rest of the world from its own viewpoint, i.e., measures, feels, forms, according to its own force."[40] And, at *WP* 647, the following: "The influence of 'external circumstances' is overestimated by Darwin to a ridiculous extent: the essential thing in the life process is precisely the tremendous shaping, form-creating force working from within which *utilizes* and *exploits* 'external circumstances.' "[41] Let's take a closer look at Nietzsche's positive understanding of teleology.

2. Teleology: From the Mechanistic View to the "Dynamic Interpretation of the World"

Though the conventional wisdom is that Darwin exploded teleology and embraced mechanism, his views on teleology are not so simple. Moore notes that

> Darwin's views on progress and teleology were ambivalent. . . . Darwin did believe in evolutionary progress: evolution was for him progressive in the sense that it pushed each form toward a higher level of organization within the context of its own peculiar kind of structure, with the result that its descendents were better prepared than their ancestors to cope with particular conditions of existence.[42]

Robert Richards, keen to save Darwin from the (neo-)Darwinists and to show how deeply his view of nature was influenced by romantics such as Humboldt, Goethe, and Schelling, goes even further: "[Darwin] is thought to have conceived nature not organically but mechanistically—as if he had to reach back to physics to secure the basic principles of his biology."[43] However, he continues, "[Darwin] never referred to or conceived natural selection as operating in a mechanical fashion, and the nature to which selection gave rise was perceived in its parts and in the whole as a teleologically self-organizing structure."[44] And even Daniel Dennett, arch neo-Darwinist, allows the question to be asked: "Did Darwin deal a 'death blow to teleology,' as Marx exclaimed, or did he show how 'the rational meaning' of the natural sciences was to be explained . . . thereby making a safe home in science for functional or teleological discussion?"[45]

The conceptual foundations of the teleology/mechanism debate reach back to Kant's philosophy of nature, and while I have not the space to sketch them in detail, I want to highlight a few brief points in order to situate Nietzsche's position on the matter. In this section, I take a detour beneath the realm of biology and examine Nietzsche's critique of mechanistic thought, since the latter is crucial to his naturalism. The reason for proceeding this way is that Nietzsche's biology is not, as it were, simply biological. His critique of mechanism in biology led him to critique the mechanistic view of nature as such, and this finds him wading in metaphysical waters. His alternative view of teleology, which is based on his notion of drives, must be seen in this context.

In the third critique, Kant's investigation of teleology revolves around the phenomenon of the organism. Robert Richards relays Kant's view of organisms:

> [F]or objects to be constituted organisms or as Kant also refers to them, "natural purposes," they have to meet the following criteria: their parts form reciprocal means-ends relationships; those parts come into existence and achieve a particular form for the sake of one another (through growth, maintenance, and reproduction); and the entire system has to be understood as resulting from an idea of the whole. No mere mechanism displays all of these features.[46]

Organisms present a special problem for the Kantian view of nature in the first critique because they clearly exhibit a kind of order and structure, yet their purposive behavior does not seem explainable by mechanical forces. As Richards explains,

> Natural phenomena, according to Kant, could only be scientifically and properly explained by appeal to mechanistic laws. Such laws would specify the constituent parts of some entity as the adequate causes of the arrangement of the whole—that was the very meaning, for Kant, of mechanistic cause. . . . Kant thus maintained that biology could not really be a science, but at best only a loose system of uncertain empirical regularities.[47]

Kant deems teleological judgments about nature "reflective" rather than "determinate" because they do not involve the application of a universal rule to a particular instance. The latter, in other words, have a universal and necessary structure that issues from the categories of the understanding. The former judgments, Richards writes, are "reflective" because they

> indicate two related features: 1) that a concept of the whole has to be empirically discovered by an initial examination of the parts; and 2) that such a concept is ultimately grounded not in a necessary requirement of nature—that is, in a natural law ultimately based in the categories—but rather in a necessary requirement of our reflective capacities.[48]

Since such judgments only express regularities, not necessities, they do not reflect the structure of the understanding and cannot in any sense constitute knowledge of the empirical world because they lack the form of universality and necessity. Only mathematical physics possesses this character, which means that, for Kant, biology is not really a science. He declares that since "in each particular natural discipline, one meets only so much real science therein as there is mathematics to be met," there can be "no Newton of the grass blade."[49] That is why he is led to dismiss any attempts at a non-mechanistic biology as nothing but "poetic swooning."[50]

So on the one hand, Kant banishes teleology from natural science. On the other hand, he maintains that we cannot help but understand living things in a teleological manner. But teleological principles cannot explain biological phenomena; we merely must act "as if" they do. However, Kant accepted the Newtonian view of nature as matter in motion governed by fixed mathematical laws. This is what motivates his dualism of a "kingdom of nature" and a "kingdom of ends." Evan Thompson gives an excellent summary of Kant's bind:

> Kant sees the futility of appealing to any immaterial principle of vitality outside of nature as a way of understanding the self-organized character of life. The only other option he can envision is hylozoism, the doctrine that all matter is endowed with life. But this doctrine contradicts the very nature of matter, which according to Newtonian physics is lifelessness or inertia. Unable to get beyond this dilemma, Kant retreats to the position that self-organization can only be a regulative principle of our judgment, not a constitutive principle of nature.[51]

The way to unravel Kant's bind is by going after matter and mechanism. Kant's view of matter is not consistent with his commitment to mechanism. This view, developed in the *Metaphysical Foundations of Natural Science*, showed, Richards writes, "that the analytical composition of the concept of matter was that of attractive and repulsive forces."[52] Schelling would exploit this to develop an evolutionary view of nature to oppose Kantian and Newtonian mechanism. As Richards details,

> Following Kant, Schelling proposed . . . a concept of matter that revealed it to be a dynamic equilibrium of the forces of attraction and repulsion. Even according to the usual beliefs of dogmatic science, he observed, our experience of material objects

and their qualities can occur only through the agency of forces that act on us. We can never experience even mediately material objects not expressive of force.[53]

"The qualities of matter," he adds, "thus displayed themselves as expressions of variously combined oppositional forces. In this way, organicity—the dynamic rebalancing of forces—constituted the fundamental property of all natural bodies."[54] Compare Nietzsche: "The connection between the inorganic and the organic must lie in the repelling force exercised by every atom of force."[55] "The drive to approach—and the drive to thrust something back are the bond, in both the inorganic and organic world."[56] As Schelling put it, "the organic never indeed arises, since it was already there."[57] This issues from Schelling's principle of "dynamic evolution": "One and the same principle unites inorganic and organic nature. . . . Every product that seems now fixed in nature exists only for a moment, and is in the process of continual evolution, a constant transformation, which would only seem played out at a particular stage."[58] The resonance with Nietzsche is obvious. What all of this adds up to, Richards writes, is that "Nature had to be conceived as a progressive evolution, achieving ever-new productive moments, never at rest, but striving toward perfection."[59] What this gives us, in lieu of an awkward dualism between the realm of natural necessity and that of freedom—with the frothy residue of "sublime nature" residing "beneath" the clockwork operation of the former—is a view of nature as a hierarchy of forms creatively emerging over time and governed by the same basic processes, with one level building on its predecessor, and with a general direction toward greater complexity and integration. This is an example of how the German idealists, especially Schelling, were trying to reconstruct the "great chain of being" in light of modern science. We will see below how Nietzsche recapitulates these Schellingian themes in his own attempts to offer an alternative to mechanism. The trick, for Nietzsche, is to advance this principle of organicism without adopting a mentalistic or vitalistic model, as if there were some force consciously foreseeing and designing organic structures; how to maintain creativity and dynamism at the inorganic and organic levels without substituting a metaphysical genie such as *Geist* for God (what Dennett derides as a "skyhook"). So the question becomes: How can we recover a conception of natural teleology while avoiding anthropomorphism or a kind of intelligent design theory? As Heidegger recognized in his own investigations in the ontology of life, reason is hard pressed to furnish an alternative between mechanism and anthropomorphism.

Kant's teleology has an "as if" status in the context of an envisioned mechanistic explanation that has an "is" status that corresponds to nature's empirical reality. But Nietzsche's aim is to pull the rug out from under this latter understanding of nature, to show that it is a bogus foundation, a conceptual abstraction. Indeed, Nietzsche's view of categorial schematization is actually a radicalization of Kant. He applies the same logic to the mechanistic mindset that Kant does to the teleological one: that because of the kind of creatures we are, we find it useful and necessary to interpret the world according to certain categorial schemes and that none of these schemes can represent things exactly as they are. We represent the world as ordered in accord with our own practical needs. As he writes,

> In order to sustain the theory of a mechanistic world . . . we always have to stipulate to what extent we are employing two fictions: the concept of motion (taken from our sense language) and the concept of the atom (= unity, deriving from our psychical "experience"): the mechanistic theory presupposes a sense prejudice and a psychological prejudice. . . . The mechanistic world is imagined as only sight and touch imagine a world (as "moved")—so as to be calculable—thus causal unities are invented, "things" (atoms) whose effect remains constant (—transference of the false concept of subject to the concept of the atom.)[60]

Nietzsche's attack on mechanistic theory is remarkably similar to Merleau-Ponty's attack on naturalism: both argue that scientific theories smuggle their concepts from sense experience without acknowledging the debt. As Evan Thompson notes, Merleau-Ponty argues that "the phenomenal domain supplies the meaning of physiological constructs" and that "naturalism needs the notion of form . . . but this notion is irreducibly phenomenal. Hence naturalism cannot explain matter, life, and mind, as long as explanation means purging nature of subjectivity and then trying to reconstitute subjectivity out of nature thus purged."[61] The common strategy here is not to attack scientific naturalism from without by "stacking" another principle or kind of being "on top" of inorganic, mechanistically governed nature—as in vitalism—but to critique it from within: by showing that it is not so stable a foundation as its proponents suppose.

This apparently antirealist view need not mean that concepts are purely arbitrary; their roots lie in sense experience and psychic drives. It means merely that they are to some degree formed by and selected from particular perspectives and that the explanations they inform will always be partial.

As Nietzsche notes, "We may venture to speak of atoms and monads in a relative sense; and it is certain that the smallest world is the most durable."[62] Flux or chaos is not pure; there is always a relative degree of stability no matter the entity in question. In fact, this notion evokes Heidegger's view of the "law of the earth" discussed in chapter 5: there is some order in the nonhuman world, though it remains opaque to us.

Much like phenomenology would attempt later on, Nietzsche's psychology is meant to be fundamental in that it is means to deflate the ontological pretensions of the natural sciences and trace their posits back to the constitutive activity of the mind. Though Nietzsche does not use the language of intentionality, his notion of drives is always drives *toward*, and he thinks these are our primary data. Moreover, his view should not be seen merely as a precursor to an "evolutionary psychology" that reduces all higher-order capacities to biological processes and rests on a scientific naturalism. For Nietzsche, psychology is not intended to apply merely to humans, but to the drives that constitute all things. When Nietzsche refers to psychology as the "doctrine of the morphology and development of the will to power," he is not restricting its scope to human beings. In attempting to explode the foundations of mechanistic science, Nietzsche was trying to combat what Heidegger would label decades later as the "tyranny of physics and chemistry" over biology. It should be noted that his concern is largely motivated by a serious problem: causal interaction between fundamentally different kinds of being. He often asserts, for example: "There is no other kind of causality than that of will upon will. Not explained mechanistically."[63] He is also concerned about the problem of emergence: how could life emerge from a purely mechanistic order if there were not already some degree of the same dynamism that constitutes the former already in the latter?

That he was concerned with such issues and that his views on nature and life attempt, however provisionally, to answer them indicate that Nietzsche's approach belongs to a certain family of related views. Heidegger's early, neo-Aristotelian ontology of life, the biophenomenological thinkers such as Jonas and Scheler (who was deeply influenced by Nietzsche), and process thought all aim to articulate similar visions of nature as value-laden and possessing hierarchical strata of organization. To further flesh out the kind of nonreductive naturalism I see both Nietzsche and Heidegger grappling toward, I want to elaborate Jonas's and Evan Thompson's positions; Jonas because he extended Heidegger's existential analytic to the region of living being as a way to displace scientific naturalism, and Thompson because he builds on Jonas's phenomenology of life and, though his notion of autopoeisis, challenges neo-Darwinism, the chief avatar of reductive naturalism today.

A. HANS JONAS'S AND EVAN THOMPSON'S PHILOSOPHIES OF LIFE

While earlier phenomenologists tended to focus on human intentionality, Jonas attempted a phenomenology of life or what he calls "an existential interpretation of biological facts." This involved an expansion of aspects of Heidegger's phenomenology of human Dasein to nonhuman organisms. The purpose of Jonas's account is to recover a vision of humans belonging to a natural cosmos that both is the source of value and, unlike premodern views of nature, incorporates a theory of evolution and is guided by an ontological principle of dynamic becoming rather than static being.

However, Jonas was uneasy about the metaphysical neutrality assumed by phenomenology; he was concerned with a philosophy of nature that comprises humans and nonhumans and a philosophy of life that comprises mind and organism. Convinced that a careful analysis of the phenomenon of life exposes the bankruptcy of the two reigning modern metaphysics, dualism and materialism, Jonas thought it incumbent on philosophy to provide an alternative—a naturalistic one, to be sure, but not what we today would call scientific naturalism.

I want to sketch two aspects of Jonas's view that I find congruent with the naturalism I am developing out of Heidegger and Nietzsche: the story he tells about the geneaology and incoherence of the modern view of life and his account of the essential features of life. Jonas's target is the modern view of nature in general and life in particular, and his goal is to expose and extirpate the fundamental assumptions that obscure and distort the phenomenon of life. In a nutshell, the distortion is that life, the living, the region of organic being, is denied an autonomous place in both philosophy and science: in philosophy, life is divorced from mind; in science, life is reduced to matter. This distortion warps our views of both our own humanity and of nature itself and is shared, Jonas thinks, by philosophies as disparate as existentialism and scientific materialism. This is what makes the prejudice so powerful and so hard to see: it is a fundamental concept of modern metaphysics, part of the background of its discourse.

Jonas regards dualism as "the most momentous phase in the history of thought."[64] In the predualistic phase, life was not a pressing philosophical problem, since the universe was understood to be intelligible in a more than materialistic sense, and life was assumed to have a natural place within something like a Great Chain of Being, a cosmic hierarchy of matter, life, and mind. In the postdualistic phase, however, what was formerly the least intelligible order of being, matter, becomes the most and, indeed, only intelligible order in nature, and life comes to be explained away as a "subtle hoax of matter":

> The discovery of the separate spheres of spirit and matter . . . created forever a new theoretical situation. From the hard-won observation that there can be matter without spirit, dualism inferred the unobserved reverse that spirit can be without matter. . . . In the postdualistic situation there are on principle not one but two possibilities of monism, represented by modern materialism and modern idealism respectively: they both presuppose the ontological polarization which dualism had generated, and either takes its stand in one of the two poles, to comprehend from this vantage point the whole of reality.[65]

What dualism achieved, in other words, was a dramatic change in the ontological default settings, especially in the being of nature.

Materialism inherited an estate that it did not build: namely, the view of nature as pure physical extension governed by mechanical principles to be comprehended through mathematical description. The setting up of this natural order was sanctioned by the preservation of a spiritual or mental order, but materialism attempts to chip away at and absorb its counterpart, and ironically, Jonas alleges, this is what leads to its downfall. As Jonas writes,

> materialism is the more interesting and more serious variant of modern ontology than idealism. For among the totality of its objects—bodies in general—materialism lets itself in earnest also encounter the living body; and since it is bound to subject it, too, to its principles it exposes itself to the real ontological test and with it to the risk of failure: it gives itself the opportunity of knocking against its limit—and there against the ontological problem.[66]

The problem is the ontological status of life.

The mechanistic approach to nature in general and life in particular was motored by a ban on anthropomorphism. However, the ban on anthropomorphism that got the materialistic project off the ground is easier to sustain when there is still a metaphysical gap recognized between the physical and the mental. But Jonas suggests that once natural science, via the theory of evolution, secured the continuity of animal and human life, a paradox emerged: nature is understood to be alien to value, purpose, and freedom, yet humans, now understood as parts and products of terrestrial evolution, exhibit such qualities. It no longer seems acceptable to keep the two orders at arms length, so we are led either to revoke or weaken the ban on anthropomorphism and admit the evidence of our own experience as living bodies and of living beings, or double down on it and, in order to be consistent, refrain from anthropomorphism—*even in regard to human beings*. The point is that the ban itself, premised on the metaphysical separation of humanity and nature, ceases to be a sound methodological principle once that metaphysical separation has been erased.

> With evolution, Jonas writes, the continuity of descent now established between man and the animal world made it impossible any longer to regard his mind and mental phenomena as such as the abrupt ingression of an ontologically foreign principle. . . . With the last citadel of dualism there also fell the isolation of man, and his own evidence became available again for the interpretation of that to which he belongs. . . . If man was the relative of animals, then animals were the relatives of man and in degrees bearers of that inwardness of which man, the most advanced of their kin, is conscious of himself.[67]

In other words, despite itself, the conclusion of Darwinian theory was a spur for rethinking the materialism that made it possible, and considering the possibility that aspects of human being—value, inwardness, freedom, purpose—were present, in diminished degrees, in life itself. So Jonas thinks that what from one perspective is Darwinism's erosion of traditional ontology is, from another, the prelude to a better ontology, one in which dynamic becoming, not static being, is the primary ontological principle. Jonas's point is that the default settings of contemporary ontology cannot pick up the frequency of life; or if they can, they only distort the signal. But an investigation of the phenomenon of life can lead us to suspend and perhaps replace those default settings.

Jonas calls his own view "an existential interpretation of biological facts." He seeks to show that aspects of Heidegger's account of human Dasein, such as transcendence, world, and care, can be attributed to living things as well. Life has transcendence both "outward" toward its environment and "forward" toward its future. It is able and driven to reach out beyond itself to appropriate matter in its environment in order to sustain itself; thus it stands in a position of "needful freedom" toward a world. Life also moves within a temporal horizon, one that upsets the "external linear time-pattern of antecedent and consequent, involving the causal dominance of the past." The temporal structure of life, Jonas argues, suggests a teleological interpretation. Life, he says,

> is essentially also what it is going to be and just becoming: in its case, the extensive order of past and future is intensively reversed. This is the root of the teleological or finalistic nature of life: finalism is in the first place a dynamic character of a certain mode of existence, coincident with the freedom and identity of form in relation to matter, and only in the second place a fact of structure or physical organization.[68]

Jonas's argument seems to be that, given, first, the evidence of careful attention to the way organisms actually operate and, second, the inadequacy of a materialist model for explaining that behavior, this "existentialist interpretation" seems eminently reasonable and should not be dismissed as "anthropomorphic."

Moreover, life has its own autonomy. Evan Thompson echoes this view with his notion of autopoiesis. As he explains,

> The paradigm [of an autonomous system] is a living cell. The constituent processes in this case are chemical; their recursive interdependence takes the form of a self-producing, metabolic network that also produces its own membrane; and this network constitutes the system as a unity in the biochemical domain and determines a domain of possible interactions with the environment. This kind of autonomy in the biochemical domain is known as autopoiesis.[69]

Thompson insists further that all living things are sense-making and possess a kind of intentionality; "an autopoietic system always has to make sense of the world so as to remain viable. Sense-making changes the physiochemical world into an environment of significance and valence, creating an Umwelt for the system. Sense-making, Francisco Varela maintains, is none other than intentionality in its minimal and original biological form."[70] Two contentious points here are that Jonas and Thompson hold that these features, such as autonomy and intentionality, are constitutive principles of living things and, moreover, that the meaning of intentionality at, for example, the cellular level, is unclear. As such, they appear to pass over from phenomenology to metaphysics. As Jonas insists,

> in living things, nature springs an ontological surprise in which the world-accident of terrestrial conditions brings to light an entirely new possibility of being: systems of matter that are unities of a manifold, not in virtue of a synthesizing perception whose object they happen to be . . . but in virtue of themselves, for the sake of themselves. . . . Here wholeness is self-integrating in active performance, and form for once is the cause rather than the result of the material collections in which it successively subsists. . . . this alone yields the ontological concept of an individual as against a merely phenomenological one.[71]

Jonas's main point is that if the autonomy of the living and its analogical connection to the mental cannot be secured, then, whatever the phenomenologist's claims to metaphysical neutrality, so long as he does not grant life ontological status, he is tacitly endorsing a kind of dualism. Instead, Jonas supports the kind of continuum view that I discussed earlier in connection with Heidegger's early work on Aristotle.

Jonas's argument for the continuity of life and mind is powered by his proposition that "life can only be known by life." We have a kind of "preunderstanding of life" that functions much like what Gadamer calls an "enabling horizon"—a means of access that opens up a sphere of phenomena, not an anthropomorphic bias that we should—or, indeed, that we are even able to—abandon. Unlike neo-Darwinian biology, Jonas ascribes a kind of interiority and intentionality to organisms; but how is this inference justified, given that we do not have "access" to, for example, what it is like to be an amoeba? Following Jonas's lead, Thompson appeals to a combination of biological and phenomenological evidence: "[W]e can, through the evidence of our own experience and the Darwinian evidence of the continuity of life, view inwardness and purposiveness as proper to living being":

> 1) To account for certain observable phenomena, we need the concepts of organism . . . and autopoiesis. 2) The source for the meaning of these concepts is the lived body, our original experience of our own bodily existence. 3) These concepts and the biological accounts in which they figure are not derivable from some observer-independent . . . objective, physico-chemical

> description, as the physicalist myth of science would have us believe. To make the link from matter to life and mind, from physics to biology and psychology, we need concepts such as organism and autopoiesis, but these concepts are available only to a bodily subject with firsthand experience of its own bodily life.[72]

Here, our tendency to impute something like mind and purpose to living things is not seen as a limitation or a liability of our human way of seeing, as in Kant, but as an enabling condition and means of access to life.

For Jonas, however, the evidence of our own living bodies is a gateway to the polarity of life—and, possibly, being—as such. He seems to think that, based on the evidence of our own and other, nonhuman organisms, we are permitted to make the extrapolation that inwardness and transcendence go, if not all the way down, at least down to the first stirrings of life. However, Jonas resists the speculative lure of panpsychism and does not follow, for example, Whitehead in imputing inwardness to nonliving things; he concedes that this is a "transference from life" that is "frankly speculative."

At the same time, however, Jonas seems sympathetic to Whitehead's metaphysical view of nature as an immanently creative, self-transcending, evolving cosmos. What he finds attractive is that Whitehead's view is evolutionary through and through. "Whatever their success so far, all contemporary revisions of traditional ontology indeed start, almost axiomatically, from the conception of being as becoming, and in the phenomenon of cosmic evolution look for the key to a possible stand beyond the old alternatives."[73] The great challenge for ontology, in other words, is to incorporate the notion of evolution without accepting its untenable materialist interpretation; exactly Nietzsche's task in the wake of the nihilism he thought Darwinism implied. This involves the revision of a number of traditional concepts, many of them Aristotelian, as we saw above with teleology. As Jonas explains toward the outset,

> Aristotle read [the hierarchy of forms] in the given record of the organic realm with no resort to evolution. . . . The terms on which his august example may be resumed in our time will be different from his, but the idea of stratification, of the progressive superposition of levels, with the dependence of each higher on the lower, the retention of all the lower in the higher, will still be found indispensible.[74]

Though these notions of ascending levels of formal complexity and increasing degrees of inwardness are suggested by our experience of living nature, they are not supplied; that is a blank that we fill in. Nevertheless, it is an extrapolation that for Jonas is well supported by the data at hand. We might differ in how we interpret that scale, but its reinstatement, against the grain of the modern view of life and nature, is essential.

In sum, though Jonas does not explicitly spell it out, his view seems to be that the task of contemporary ontology is to integrate the evidence of our own experience of nature, afforded by phenomenology, and that of natural science, especially evolutionary theory, into a broader framework. Note that Jonas' motivation for working out an alternative naturalism is ultimately ethical. In his view, the chief problem with the modern understanding of life and nature is nihilism. If values are merely understood as projected onto nature by humans, then values have no objective ground. As such, Jonas' project points toward what was traditionally called the metaphysical ground of ethics. He seeks "a principle of ethics which is ultimately grounded neither in the autonomy of the self nor the needs of the community, but in an objective assignment by the nature of things."[75] Without this, he thinks we are stuck with the problem of nihilism, or as he puts it, "anthropological acosmism," with humans as valuers in an inherently value-less world. His aim is to recover a conception of natural value rooted in the "breadth of being" in order to overcome nihilism.

I want to mention a few more points from Evan Thompson's autopoietic view of the organism as a self-producing and self-regulating system that enacts, brings forth, or constitutes a meaningful environment, since I think they complement Nietzsche's view. On three points—causality, matter, and teleology—Thompson explains why strains in contemporary theoretical biology are pointing away from mechanism, neo-Darwinism, and Kant's restriction of teleology to the status of a regulative principle; Nietzsche's views in many ways prefigure these developments. First, Thompson explains why Kant's bind is "no longer compelling" largely because of progress in science:

> [T]wo kinds of scientific advances have been decisive. The first advance is the detailed mapping of molecular systems of self-production within living cells. We are now able to comprehend many of the ways in which genetic and enzymatic systems within a cell reciprocally produce one another. The second advance is the invention of mathematical concepts and techniques for analyzing self-organization in nonlinear dynamic systems. . . . Many

scientists now believe these are necessary principles of biological self-organization.[76]

Nietzsche's view of organisms as relatively stable configurations of drives that in some sense produce themselves, and that even cells cannot be understood mechanistically, prefigures this view.

Second, Kant's view of matter is outdated. As Thompson explains,

> Our conception of matter as essentially equivalent to energy and as having the potential for self-organization at numerous spatio-temporal scales is far from the classical Newtonian worldview. In particular, the physics of thermodynamically open systems combined with the chemistry and biology of self-organizing systems provides another option that is not available to Kant: life is an emergent order of nature that results from certain morphodynamical principles, specifically those of autopoiesis.[77]

Though he did not have access to the science we do, it seems that Nietzsche's basic intuition that mechanism would be superseded by a "dynamic interpretation of the world" centered on quanta of energy was generally correct.

Finally, Thompson explains why the autopoietic view underwrites a naturalized teleology or "immanent purposiveness":

> The first mode of purposiveness is identity: autopoiesis entails the production and maintenance of a dynamic entity in the face of material change. The second mode of purposiveness is sense-making: an autopoietic system always has to make sense of the world so as to remain viable. Sense-making changes the physiochemical world into an environment of significance and valence, creating an *Umwelt* for the system. Sense-making, Varela maintains, is none other than intentionality in its minimal and original biological form.[78]

Here I think we have something very much like Richardson's notion of "thin intentionality." And this gets at Nietzsche's connection of interpretations, drives, and values: each organism interprets its natural environment based on its distinctive drives. And each interpretation, the way in which an organism constitutes its environment, is evaluative. As Varela puts it, "[sense-making] lays a new grid over the world: a ubiquitous scale of value."[79] The key here is that Kant could envision teleology only in a transcendent, "top-down" fashion—that an intelligent mind designed the end toward which a thing develops—rather than in a "bottom-up" fashion, as emerging through the interactions both within an organism and between it and its environment. So the picture that begins to emerge here is that Nietzsche follows Kant in rejecting transcendent teleology but parts from Kant in embracing an immanent teleology.

To sum up the various strands developed in this subsection: Nietzsche was trying to build a bridge between the inorganic and the organic, on the one hand, and biology and psychology, on the other to provide an interpretation of the world that could integrate matter, life, and mind. This interpretation rejected mechanism in favor of an alternative teleology, the cornerstone of which is his conception of drives.

3. Drives

Nietzsche's position is that all living beings are composed of what he calls drives, which are basically synonymous with "wills." The will to power is nothing else than these drives; it is not a supreme being or a supernatural force "behind" the drives that compose actual beings. Richardson explains how the notion of drives underwrites Nietzsche's teleology:

> Nietzsche doesn't really give up explaining by ends. His key notions of wills, drives, and values all involve directedness or aiming. . . . The key point Nietzsche takes from Darwin is a different model for teleology, which he extends and applies. It's this core Darwinian insight that lets him naturalize his wills and drives: goals can be set into organisms—they can be designed for certain outcomes—by processes that don't at all "represent" or "foresee" those outcomes. At issue, in particular, [is] how Nietzsche can attribute the end-directed character he clearly does to these drives and wills, without illicitly anthropomorphizing an implausible mentality into them.[80]

The question is about the primary engine of evolution: Is it the will to power that actively directs the development of organisms (i.e., do they direct their own development?), or does it merely produce chance variations that provide the raw material for natural selection? The former is metaphysical and vitalistic (a "skyhook"), while the latter is naturalistic (a "crane," in Dennett's parlance). Richardson holds that the former is Nietzsche's "dominant" view and that the latter is his "recessive" view. He attempts to pare away the metaphysical baggage and argues that much of what Nietzsche says about drives can be understood naturalistically through a notion of " 'thin intentionality' not dependent on mentality."[81] I find Richardson's view too acquiescent before neo-Darwinian orthodoxy, but let us look at his presentation in order to get a better handle on Nietzsche's view of drives.

First of all, let us look at some passages in which Nietzsche attacks the mentalistic view of teleology. In one of the "Against Darwinism" sections in *WP*, after faulting Darwin for relying too heavily on "external circumstances" to the neglect of the "shaping, form-creating force working from within which utilizes and exploits 'external circumstances,' " he is careful to note that "the new forms molded from within are *not formed with an end in view*; but in the struggle of the parts a new form is not left long without being related to a partial usefulness and then, according to its use, develops itself more and more completely."[82] Nietzsche also attacks this view as it applies to human consciousness. At *WP* 707, he describes how the error of mentalism arises and subsequently distorts our view of life:

> In relation to the vastness and multiplicity of collaboration and mutual opposition encountered in the life of every organism, the conscious world of feelings, intentions, and valuations is a small section. We have no right whatever to posit this piece of consciousness as the aim and wherefore of this total phenomenon of life: becoming conscious is obviously only one more means toward the unfolding and extension of the power of life. . . . This is my basic objection to all philosophic-moralistic cosmo- and theodices, to all wherefores and highest values in philosophy and theology hitherto. One kind of means has been misunderstood as an end; conversely, life and the enhancement of its power have been debased to a means. . . . The fundamental mistake is simply that, instead of understanding consciousness as a tool and particular aspect of the total life, we posit it as the standard and the condition of life that is of supreme value.[83]

Again, what Nietzsche is saying is that the real problem is the relation between mind and life. The fact that it has been framed as the "mind-body" problem is a symptom of their dissociation. Note that Nietzsche is not reducing the level of consciousness to biological drive-life and adopting a kind of epiphenomenalism. Though genealogy is supposed to show how consciousness emerged out of, depends on, and is influenced by subconscious drives, and is itself a constellation of drives, that does not rob it of causal power; otherwise, his calls for self-overcoming and "passionate mastery of the passions" would make no sense. Consciousness is but the most recently emerged, most complex, and most fragile aspect of our being, and while it realizes value, it must be integrated with the other strata, harmonized with the other drives. There is a pattern at work here that we're going to see again and again: when any drive or set of drives attempts to destroy and drive out those on which it depends and from which it emerged, it sabotages itself. The important point here is that Nietzsche not only avoids imputing consciousness to nonhuman life forms, but he also critiques it as a model for understanding human life. Moreover, this strain in Nietzsche's thought—namely, resistance to postulating metaphysical principles to explain biological phenomena—should make us wary of branding him a vitalist.

At the same time, there is an abundance of passages that appear to contradict this view; there is simply no getting around the fact that at times Nietzsche makes unmistakably vitalist and/or panpsychist claims. Nietzsche sometimes speaks as if cells are "self-conscious," and as though there were "a mass of consciousness and wills in every complex organic being."[84] Indeed, as Moore has shown, Nietzsche was surely influenced by vitalist biologists such as Michael Foster, who attributed volition to primitive organisms such as amoebas and hydras. One of Nietzsche's signature views of personality and organic unity—that it is actually an aggregate or coordination of many subwills—seems to have come from Foster. As Moore notes, Nietzsche "even characterizes the inorganic world as 'consciousness without individuality'; all that differentiates the organic from the inorganic world is that the former has developed a degree of subjectivity, a 'perspective of egoism.'"[85] When he speaks like this, Nietzsche does seem to attribute consciousness to all things. Moore is correct to claim that for Nietzsche, "[c]onsciousness is not the exclusive prerogative of human beings, or even of highly developed organisms, but is rather an amplification, an evolution of patterns and processes present in the organic world as well as the most basic organic material."[86] On these occasions, I think we have to concede that Nietzsche is at best imprecise and at worst sloppily anthropomorphic, and insofar as he embraces views that directly contradict established biological science—for which vitalism has gone the way of phlogiston—his views must be rejected.

But I think the contradictions in his views on these issues may have to do, in part, with ambiguity over the meaning of "consciousness." If by consciousness Nietzsche means what we understand as self-awareness, a rich interior life, emotions, abstract/representational thinking, sophisticated cognitive abilities, and so on, then attributing it to all things is surely absurd. But if by consciousness we mean something more minimal, an inner dynamism, a basic directedness toward the world, a perspective capable of registering and responding to a limited range of stimuli or phenomena, and so forth, then that is, at the least, less implausible. Note how similar this is to Heidegger' pre-*Being and Time* view, namely, that all living things are in some way directed toward the world. Heidegger's earlier reading of the *De Anima*, where he saw soul as an ontological, not a merely psychological, principle, has strong connections to Nietzsche's view. In his magisterial study of Nietzsche's psychology, Graham Parkes picks up on the affinity with Aristotle: "[I]t appears that for Nietzsche the answer to the question of what the soul is like is that ultimately it is like everything. . . . Not only natural worlds but the worlds of human community move and have their being within as well as without. As Aristotle said, the soul is in a way all things, and so the boundaries between inner and outer are dissolved."[87]

Richardson's notion of "thin intentionality" is a start for making this understanding of the will to power biologically palatable. He attempts to show that Nietzsche understands the telic character of drives to be naturally selected. As such, he fashions the following definition of will to power: "Will to power is a disposition to cause a certain result, i.e., power, and past such results caused (produced) this disposition."[88] The telic character of a present drive can only be understood by reference to its past: "The meaning of a drive today is a layering of the functions it was serially selected for, in becoming what it is."[89] Richardson is modeling his functionalist account of teleology on that of philosopher of biology Robert Brandon, whose formula is as follows: "[T]rait A's existence is explained in terms of what A does. More fully, A's existence is explained in terms of past instances of A; but not just any effects: we cite only those effects relevant to the adaptedness of possessors of A."[90] So the explanatory end here is reproductive fitness. And the function or purpose of a drive is not intentionally determined by the individual or organism, but by the design history operating "behind its back." By regarding wills to power as selected, Richards thinks, we can cast Nietzsche's views in their "least metaphysical" form: "[T]he drives that have best served reproductive success and that dominate the drive economy of most organisms are drives whose goals involve some kind of control, either over other organisms, or over other drives in the same organism."[91] The "structural end" of selection is still preservation, but power enhancement is "plastic" toward this goal. In other words, natural selection is always in the driver's seat, but drives that aim at power-enhancement rather than mere preservation or reproductive success actually conduce, accidentally and not intentionally, toward the latter. In this way, Richardson thinks, Nietzsche's views are consistent with neo-Darwinism.

Despite Richardson's careful reconstruction and his claim that there is "overwhelming evidence that Nietzsche considers drives to be (at least in large part) products of natural selection,"[92] he concedes that his naturalizing strategy only goes so far because it reduces will to power to mere "mutability": "All the richer ways Nietzsche thinks organisms are structured 'for self-overcoming' collapse when we remove his Lamarckian support. Without it, organisms can't be designed to overcome themselves ('in their lives'), as a way to improve the species. The only place design for evolution occurs is in the copying process."[93] Richardson thinks that we must reject Nietzsche's view as a metaphysics or "power ontology" and prune it down to a "power biology." But this still leaves the problem of just how the organic and inorganic realms are related, and if we punt on the ontological question, the vacuum is going to be filled by the scientific naturalist, who is going to insist on mechanism. Yet, as I have tried to show, this generates the problem of nihilism, of how we can have value, meaning, and purpose in a world that does not admit them. And it is the reason, I think, that Nietzsche resisted a mechanistic account of life and, at times, threw in for vitalism. He wanted to hew as closely as possible to a naturalistic philosophy but was caught in the bind Heidegger was later to identify: "[H]uman analysis practically runs out of alternatives when it rejects mechanistic views . . . as firmly as it avoids anthropomorphic interpretation."[94]

I think Richardson is too acquiescent before neo-Darwinian orthodoxy, which automatically dismisses nonmechanistic accounts as so much "spooky" vitalism. As John Cobb Jr., writes, there are plenty of empirical reasons to doubt this orthodoxy:

> [T]here is evidence for the importance of epigenetic factors within the cell. There is evidence for lateral transfer of genes and even complete genomes as well as mutation. There is evidence that what happens to the life-form influences the selection of genes. There is evidence that the activity of these life-forms affects the environment in ways important to what is selected. . . . The "nothing-but" formulations of leading neo-Darwinists are profoundly misleading and destructively restrictive.[95]

Indeed, neo-Lamarckian approaches to genetic variation have gained support in recent years. While "primitive Lamarckism"—the proverbial giraffe "willing" to grow its neck—has been discredited, there is evidence for the role of epigenetic and behavioral factors in genetic variation. There is no doubt that Nietzsche sometimes appears to speak in support of "primitive Lamarckism," but the most charitable reading is that he correctly gestures, however crudely, toward a nonreductive view of evolution in which the life and behavior of the organism play a part in its development and legacy. Famed biologist Richard Lewontin, as John Greene explains, argued against viewing the organism as the passive plaything of genetic and environmental factors and that "biologists should recognize that the organism constructs its environment by its activities, and that the effective environment consists of those aspects of the external world that are relevant to those constructive activities."[96]

So despite his vitalist leanings, Nietzsche pointed, incompletely and inconsistently, toward nonreductive approaches to biology that would gain greater credence in the twentieth century. Cobb argues that

> the activity of the organism plays a role in determining 1) the character of the natural environment, 2) which genetic mutations will be selected, 3) how lateral transfer of genes occurs, and 4) the occurrence of epigenetic influences on evolution. The neglect of these factors . . . suggests that the materialist assumptions so widely operative in scientific theory make it difficult for biologists to give full weight to some aspects of the empirical evidence.[97]

Where neo-Darwinists hold that "the significant causal relations [in evolution] are unidirectional—from the gene to the organism and from the environment to the organism,"[98] Cobb and others (and Nietzsche) hold that evolution is more complex and that organisms are, to some extent, agents in the process. Opponents of neo-Darwinism simply ask for its proponents to either come clean and espouse and defend a metaphysical materialism—which most, concerned to preserve biology as an autonomous field of study, may not want to do—or to acknowledge and abandon the materialistic metaphysics assumed in their theory and allow for nonmechanistic factors. Cobb shows why, for Whitehead and process thinkers, the questions of progress and value are so important in evolutionary theory:

> For something to have value in itself, it must have existence or reality for itself. . . . Whiteheadians believe that all the unitary entities that make up the world have reality in themselves as well as for others. Therefore, all have some intrinsic value. . . . We believe that animals with central nervous systems have far richer

experience than unicellular organisms or the cells that make up plants and animals, although these cells are not valueless. For this reason we believe that evolution began with entities of very modest intrinsic value and has produced creatures of much greater value. Overall, therefore, there has been enormous progress. From the perspective of process thinkers, then, the inability of thorough-going neo-Darwinians to speak of progress is a weakness. It leads them in the direction of nihilism.[99]

So again, we see that one of the fundamental problems in a mechanistic, materialist biology, stretching from Kant to contemporary neo-Darwinism, is its tacit nihilism. While Nietzsche's view is not Whitehead's, and despite his attacks on modern progressivism, he, too, appears to hold that there is in some sense a hierarchical, progressive dimension to the will to power through which higher values are realized. To make this clear, we need to look at two more aspects of his biology: values and selection.

4. Values

Nietzsche is sometimes regarded as having a "projectionist" thesis about values: namely, that values are not objective in any sense but are merely subjective human projections motivated by practical needs and interests. There is no doubt that he sometimes speaks this way. To cite but a few examples: "Whatever has value in the current world, has it not in itself, from nature—nature is always valueless—but one has once given it a value, as a gift, and we were those givers and gifters!"[100] "The human first laid values into things, to preserve himself,—he first created a sense for things, a human-sense!"[101] These passages would appear to vitiate attempts to pin him to any theory of natural value, since values would merely be imputed to objects but would in no way be metaphysically anchored in them.

However, we should not too hastily take remarks such as these at face value. For one thing, Nietzsche often hyperbolizes in order to provoke; since he is criticizing the status quo, he tends to overcompensate by making his alternative sound more extreme than it actually is. Second, it is different to say "nature is valueless" than it is to say "natural things value or have value." Since Nietzsche thinks that reality is composed of perspectives, drives, or wills to power, "nature in itself" is just an abstraction; there is no "thing" called nature, only the various perspectives that compose it. Third—and most important—it is beyond dispute that he held valuing to be an inherent activity of all living things, saying, "Valuations lie in all functions of the organic being."[102] He also says, "'Higher' and 'lower,' the selecting of the more important, more useful, more pressing arises already in the lowest organisms. 'Alive': that means already valuing."[103] His project to "naturalize values" is not so much to show how all human valuing is empty, has no referent, as to show that all values were creative responses to life conditions that, over time, became habituated into social norms and hypostasized as cosmic constants, and that values only exist *as* valued. As Richardson explains,

> A first important way in which he "naturalizes values" is precisely by insisting on their dependence, as contents, on those activities of valuing—so putting them back into their natural setting. A value is always "for" a valuing; it is an intentional object of that valuing and ontologically dependent on it. There can only be goods, as posited by a valuing viewpoint.[104]

There are no "values in themselves" or entities with values as "properties"—there is only valuing activity. For Nietzsche, activity as such is already evaluative. Valuing, for him, is not merely aesthetic or moral, but ontological; it is not merely something beings sometimes do and sometimes don't—it is something they are. If beings are composed of nothing but drives, and all drives value, then beings value intrinsically; the drives that dominate will determine what the being values. So, contrary to the projectionist thesis, Nietzsche does reserve a place for the "reality" of values. Richardson clarifies this:

> [T]he dependence of values on valuings does not imply that there *are no* values; rather, it tells us what they are. There are values in the world . . . precisely because valuers have put them there, by their aims and intents. As I will put this point, he thinks that values are *real* . . . but not *objective* (i.e., values always exist for a "subject"—construed very broadly to include the drive or will he finds in all organisms).[105]

Again, this follows from Nietzsche's critique of the mechanistic view of nature: when reality is no longer defined as externally related objects of matter in motion, casting value in terms of the valuation of subjects is no longer "merely" (humanly) subjective, since subjectivity, in some form, is recognized as a constitutive feature of reality.

But this should not be mistaken to mean that valuing is necessarily or even primarily cognitive. Richardson explains further why Nietzsche's value theory is not anthropocentric, strengthening the connection with evolution:

> [Nietzsche] thinks we commonly suppose that goods and values are confined to an autonomous human and psychological domain—that we alone have values by virtue of our singular mentality. . . . But according to Nietzsche, we not only share values with other creatures, but even in us the really effective or influential values are not those conscious ones, but values we have, as it were, through the plant and animal in us. Values are built into our bodies, and their conscious and linguistic expression is something quite secondary.[106]

But how are we then to understand the directed-ness of valuing, if not in cognitive terms? This is where Richardson's notion of "thin intentionality" comes in. Drives, to paraphrase his definition, are plastic dispositions to behavior. They are not blind mechanisms, as in behaviorism. This notion of plasticity is what Merleau-Ponty attempted to capture in his conception of "structure" in his analysis of animal behavior in *The Structure of Behavior*, in which he draws on Uexküll's notions of *Bauplan* and *Umwelt*.[107] As Evan Thompson explains,

> [For Merleau-Ponty,] to say that stimuli play the role of occasions rather than cause is to say that they act as triggering conditions but not as efficient causes. To say that the organism's reaction depends on the vital significance of the stimulus is to say that the informational stimulus is not equivalent to the physical stimulus. . . . Something acquires meaning for the organism to the extent that it relates (either positively or negatively) to the norm of the maintenance of the organism's integrity. . . . Behavior is, as it were, dialogical and expresses meaning-constitution rather than information processing.[108]

Plasticity preserves the organism's capacity to respond creatively to environmental pressures, and responses that are naturally selected constitute—but do not exhaust—that organism's "good." As Nietzsche writes, "Every drive is the drive to 'something good,' seen from that standpoint."[109] So, Richardson says, "The definition of 'value' is equally a definition of 'good': a drive's goods are precisely its goals—the outcomes it was selected to bring about."[110] Again, we need not take Nietzsche's qualifier, "from that standpoint," to cancel the "real" goodness realized by the drive. If the selection and stabilization of a drive come to constitute a condition for the preservation and enhancement of the organisms and its species, then we can say that that is one of the constitutive goods of that thing, so long as its standpoint persists or is subsumed by another in which the drive is subordinated to (but still foundational for) others.

Finally, (again) despite the neo-Darwinian view that Darwin offered a value-free view of nature, Nietzsche and Darwin share common cause in finding value in nature. Robert Richards argues that "the usual interpretation of Darwinian nature is quite mistaken, that Darwin's conception of nature derived, via various channels, in significant measure from the German romantic movement, and that consequently, his theory functioned not to suck values out of nature but to recover them for a de-theologized nature."[111] "Darwin's nature," he asserts, "progressively produced organisms of greater value."[112] Moore also hits upon this progressive, hierarchical aspect of Darwinism, but taking the accuracy of neo-Darwinism to be axiomatic, he perceives it as a weakness: "For all Darwin's attempts to dissociate himself from the legacy of traditional biology, vestiges of the earlier, neo-Platonic concept of nature as a chain of being persist in his work. His metaphor of the tree of life . . . appears to suggest a hierarchical order of natural forms."[113] True, but Darwin gave an account of how this chain arose in real time through a natural process without divine causality, and in that sense, his view is basically aligned with Nietzsche. To be sure, Nietzsche wants the emphasis put on organisms' valuation through their activity, rather than on them as passive bearers of value, but the parallel holds. If an exceptional individual executes a creative and adaptive response to the environment, its new behavior can become an exemplar for others that, over time, gradually settles into a new structure that eventually becomes a new norm for that population or species (and alters its environment); to the extent that this new behavior preserves and enhances the life conditions of that group, it should be construed as a new valuation, an increase in power, and a kind of progress. Given Nietzsche's unrelenting support for the hierarchical perspective, it cannot be denied that he believed in a rank order in nature. Nietzsche embraces a *scala natura:* one supported not by theology or classical metaphysics, but by scientific and phenomenological findings.

5. Selection

Now we can turn to Nietzsche's broader understanding of selection and plug his biology into his two-fold project of the genealogy of values and the revaluation of values. One of Nietzsche's main tasks is to determine how the values of the present have been selected—how nihilism arises. As I discussed above, Nietzsche's conception of drives can largely be accounted for in terms of natural selection. Yet his main interest is in what this biological background can tell us about human psychology and society, and he has no doubt that when it comes to this latter sphere, natural selection is not going to do the job. As Richardson explains, "Darwinian natural selection is only a first stage in the formation of values, and a first factor in explaining them. There are other selective mechanisms besides this one."[114] He labels these "social selection" and "self selection." What this gives us is a hierarchy of three levels of selection: natural, social, and self. Each emerges out of the one that precedes it and adds a new layer of complexity. The challenge at each new level is to find a way to incorporate the drives of previous levels without letting them dominate, on the one hand, and without dominating them—in the sense of completely denying them expression—on the other.

The second level is social selection. Richardson usefully distinguishes the patterns laid down at the social level as "habits" (a kind of "second nature" that is taught), whereas those engrained at the biological level are "drives" (a kind of "first nature" that is inherited); since the latter have been replicated and reinforced over the course of millennia, while the former are relatively recent, the drives are extremely resistant to alteration. What Nietzsche calls "morality" is the attempt to uproot and dominate these drives through the inculcation of social habits. So social selection is the process of taming, domestication, and reining in natural impulses. This counters the differentiation and struggle of the natural plane with homogeneity, leveling, and herding. As Richardson points out, Nietzsche viewed social selection as an ingenious means for accelerating growth because "a much more effective way for a habit to replicate itself is by social transmission—laterally, not to descendents."[115] Note that, depending on the context, the "herd mentality" can foster or retard growth. The idea here is that, at one point in evolution, social selection is an emergent response to life conditions that fostered growth, but that, after a while, it becomes geared solely toward preserving itself and thus inhibits further enhancement. Despite its difference from natural selection, however, social selection is like natural selection in that it is not primarily driven by individuals: "Like natural selection, it is stochastic, not intentional, and not the work of any deliberate selectors."[116] Social selection is where language and consciousness emerge; for Nietzsche, both of these abbreviate the world by a system of common symbols that, over time, coordinate and organize people around a common purpose and vision of the world.[117] However, as Richardson points out, the problem is that social selection "works, to an extent, against natural selection. It works to modify behaviors designed for the organism's survival and reproduction, and to re-aim them toward goals serving a different overall end. So taught behaviors oppose or counter inherited behaviors."[118] And this tension between the unstoppable force of natural drives and the immovable object of social habits brings us full circle to what Dan Conway describes as the source of nihilism for Nietzsche: the formation of the bad conscience, the reaction resulting from the consciousness of mortality.[119] The basic conflict of drives within is negotiated by projecting a dualism without, when in reality the conflict is between two different modes of the same fundamental force, and needs to be dialectically mediated.

The idea of social selection is that each society is going to have a kind of basic framework of values that defines it; as Nietzsche puts it, a people's values are what they bow before, the common goals they are pursuing, the goods whose pursuit binds them together. So the habits and customs of a people—what is today sometimes called a "background"—are very similar to what Heidegger calls average everydayness. The individual is thrown into this social milieu, does not choose it for herself, and it subtly but powerfully shapes her outlook and engagement with the world; and indeed, Nietzsche's critique of morality is geared toward revealing the false consciousness in which we think we choose certain values, but they are actually chosen for us. But once the individual becomes conscious of the limits of this social milieu—conscious of her freedom from it—that leads to a new level, that of self-selection.

When self-selection comes on the scene, the person realizes that all values are relative in the sense that they are not anchored in any eternal, transcendent cosmic order but were at some point laid down, posited, created in response to life conditions. Self-selection is the domain of both the genealogist and the sovereign individual or over-human; indeed, these two perspectives, the negative and positive, correlate to the figures of the lion and the child in the section "On the Three Metamorphoses" from Zarathustra discussed in previous chapters, while the "camel" correlates with social selection. The crucial category here is freedom. Indeed, freedom makes genealogy possible: it is only because we are not wholly determined by biology and culture that we can engage in critique. But genealogy demands revaluation: without a positive vision with which to move forward, genealogy is just nihilism. Nietzsche did not intend the revaluation as some radically different set of values that would be unrecognizable to old perspectives. This is why, as Richardson points out, there are "two features of this value creating: 1) it is not *ex nihilo* (it remakes existing values); 2) though perhaps not compelled by the facts, his values are crucially informed by them."[120]

Just as evolution works by building one level or system on top of another, and repurposing the aims of the previous level to fit challenges on the new one, so the revaluation must not be arbitrary; it must include and incorporate the value perspectives that preceded it and enabled it to emerge in the first place. This is why, Richardson notes, revaluation "rests heavily on empirical insights: science's facts about our species' and culture's evolutionary paths, extended in each case into an individual's insights into his personal makeup."[121] Once we have become conscious of and taken responsibility for our naïve positing, we are not to rest in nihilism and refrain from valuation, but to advance values that are life-affirming—in the dual sense of enhancing human power and flourishing by integrating natural drives and social habits, and of acknowledging the perspectives and valuings of nonhuman entities. Nietzsche clearly believes, with biocentrism, that the overcoming of a narrow, humanistic perspective involves a recognition of the kinship we share with all living things—that they, too, possess something like a perspective with its own particular values, and that they embody goods with which our own is bound.

A problem arises: first, how can we speak of values as being "higher" or "better" than others if all values are merely relative to a certain perspective? Second, if all perspectives are merely "interpretations" of the world—since there is no "nature as such," there are only perspectives on perspectives—then how can one be truer than another? In other words, both questions are asking how Nietzsche can still lay claim to hierarchical language of higher and lower, better and worse, once he has made the perspectival move. Richardson sheds some light on the first question:

> Values are put into the world by all the wills (or aiming organisms) there are. So the adequacy of our own values gets judged not by whether they match the values in the things themselves (there are none), but by how they stand toward that field of competing values. . . . My values will be less partial to the extent that I somehow recognize or encompass or supersede some of those others.[122]

Nietzsche's notion of freedom or self-selection—represented by the overhuman or the sovereign individual—is characterized by a decrease in egoism and an expansion of identity beyond the merely human. This higher perspective realizes its vital origins and kinship with all things, and its moral center of gravity is thus less anthropocentric. So, as Richardson puts it, "[Nietzsche's] values are less partial and less blind than prior values—

they better recognize the whole sum of values there are, by the valuings of humans and other organisms."[123] This higher perspective is marked by inclusivity and creativity. These represent the two basic poles of valuation: preservation (inclusion of multiple perspectives) and enhancement (creation of a worldview that takes these narrower perspectives up into an integrated synthesis) of life conditions. So this value scheme is a hierarchy in which the higher, more developed, more evolved, more powerful viewpoint is less dominating, less oppressive, less anthropocentric. The capacity to recognize and take on other perspectives is a developmental achievement, and to that extent, I think Nietzsche is committed to a kind of progress, despite his occasional protests to the contrary.

Conclusion

What we can take from this chapter is that Nietzsche provides us with a view of nature that is meaning- and value-laden. He does this by liberating biology from mechanism, on the one hand, and naturalizing values, on the other. In doing so, he connects two of the major and seemingly disparate developments of the nineteenth century: the advent of biology and the rise of nihilism. An early adherent of a nonreductive naturalism, he provides us with a basis for thinking the deep continuity of the human and the living. Moreover, he offers us a view of nature as hierarchically structured, in contrast to the "flat" view bequeathed by modern science and philosophy. It recovers the depth and verticality of the traditional great chain of being, but it does so in a "bottom-up" fashion; the levels emerge progressively over time through natural processes and depend on and evolve in relation to those before them. In this way, his account can supplement Heidegger's early ontology of life, which did not engage evolutionary biology. Having laid out a naturalized form of Heidegger's and Nietzsche's thought, I turn in the final chapter to sketching how it might connect to an environmental ethic.

8

Engaging Environmental Ethics

In this final chapter, I explain how the nonreductive naturalism intimated by Heidegger and Nietzsche might provide a conceptual foundation for environmental ethics. Most environmental thinkers would probably call themselves "naturalists"—but which kind of naturalism do they have in mind? If an environmental ethic bases its view of nature on the natural sciences—scientific naturalism—it is arguably seeking for values in a valueless world and is plagued by the problem of nihilism. A nonreductive naturalism attempts this problem by anchoring value in the natural world through a phenomenological account of the organism's relation to its environment. Likewise, most contemporary phenomenologists would probably call themselves "realists"—but which kind of realism do they have in mind? I reject the agnostic posture toward metaphysics often struck by phenomenologists and hold that a nonreductive naturalism maintains a place for intentionality in the natural world. The vision intimated by Heidegger and Nietzsche (and Jonas and Thompson) reconstructs traditional views of nature as a great chain of being or *scala natura* but does so without speculative supports and in a way that is consistent with evolutionary biology. I submit that this view of nature might support a "hierarchical biocentrism," which recognizes the value of all living things while maintaining that higher, more complex forms of life embody greater value.

Having situated Heidegger in relation to environmental ethics in the first chapter, here I address the problems and promise of drawing an environmental ethic out of Nietzsche's version of naturalism. There has been relatively little debate about Nietzsche's place in environmental ethics, but the lines of that debate are clearly marked.[1] He has been framed as an anthropocentrist (Zimmerman), a humanist (Acampora), a biocentrist and deep ecologist (Hallman), and an ecocentrist (Parkes). His position is hard to pinpoint, perhaps because he was writing before there was any environmental movement and had different concerns than many environmentalists.

As Zimmerman points out, "[Nietzsche] would have criticized the kind of anti-anthropocentrism that guides much of today's environmentalism. . . . He was concerned not with biospheric nature, but rather human nature. . . . Nietzsche's major concern was how to avoid degeneration and nihilism, not how to avoid environmental destruction and ecocide."[2] On the one hand, he is very critical of humanism; on the other, he eschews the egalitarian and antihierarchical strains found in many environmental thinkers committed to notions of intrinsic value in nature. But in my view, it is *because* of Nietzsche's laserlike focus on nihilism that he can be enlisted as an environmental thinker.

I contend that Nietzsche does provide us with a basis for a theory of natural value. I review the secondary literature and submit that there are three main problems plaguing the debates. First, they tend to ignore or downplay Nietzsche's biology. The foundation excavated by Richardson, as we saw, is crucial to Nietzsche's account of natural value. Second, Nietzsche's value theory is not adequately addressed. When it is, he is misinterpreted as espousing what I referred to above as a "projectionist" theory about values, which would be highly anthropocentric. This error results from the first error: ignoring the biology. Third, most of the extant views either ignore or misunderstand Nietzsche's conception of hierarchy. I argue that when we plug Nietzsche's view of nature into the perspective of environmental ethics, his emphasis on hierarchy enables us to maintain that human life is more valuable than other life forms but that lower life forms have a different kind of value insofar as they support and enable higher forms. I connect this view to those of process ecologist David Ray Griffin and integral ecologist Michael Zimmerman, whose views, I argue, roughly parallel Nietzsche's.

I. Debating Nietzsche's Relevance for Environmental Ethics

The first attempt to situate Nietzsche within environmental ethics was made in 1991 by Max Hallman. Hallman argues that Nietzsche should be seen as an ecological or environmental thinker. Hallman bases his claim on four points: Nietzsche 1) rejects the notion of any transcendent or supernatural order, 2) rejects the human/nature dualism and criticizes anthropocentric positions, 3) has a relational rather than a substantial ontology that prefigures models of nature in ecological science, and 4) "calls for a kind of 'return to nature.'"[3] I think Hallman's main points are basically correct, but there are some problems with his view that I touch on below.

Hallman points out that Nietzsche's views about nature changed throughout his writings. Early on, in the *Birth of Tragedy* and *Untimely Meditations*, Nietzsche embraces what John Passmore dubbed a "man as creator or perfecter" model of nature. Exemplary human beings are painted as the goals of the evolutionary process and the redeemers of the warped wood of nature; philosophers, saints, and artists must "acquire power so as to aid the evolution of *physis* and to be for awhile the corrector of its follies and ineptitudes."[4] Nietzsche soon abandons this anthropocentric perspective for two main reasons: he rejects the Christian and Darwinist views of humans as the crown of nature and evolution, and Hallman writes that since Nietzsche "argues both that nonhuman species have 'knowledge,' and that human 'knowledge' is not be accorded a privileged status,"[5] he "intimates that the values or perspectives of nonhuman life forms must be taken into consideration."[6] Thus Hallman thinks that Nietzsche arrives at a kind of nonanthropocentric biocentrism or bioegalitarian view like that found in deep ecology, which critiques all notions of "hierarchy" and maintains an ontological egalitarianism or "heterarchy" (all members of an ecosystem have an equally important role to play in the system) and an ethical egalitarianism (all are equally valuable).[7]

Hallman's notion of a "return to nature" seems problematic. On the one hand, he acknowledges that Nietzsche rejects a romanticist position: he cites the passage where Nietzsche says, "Not 'return to nature. . . . Man reaches nature only after a long struggle—he never 'returns.' "[8] On the other hand, Hallman insists that Nietzsche's view hews close to deep ecology by "rejecting separateness and individuality and affirming wholeness and totality."[9] However, if the nature we are to "return to" is merely nature as understood by ecological science, and if Nietzsche thinks this leap involves the transcendence, development, and creative transformation of the human being and believes that the human is part of nature, then this implies that nature *does* exhibit some kind of transcendence.

Along similar lines, Hallman argues that

> the will to power primarily serves as a principle that explains change immanently according to certain homeostatic relations, a principle that theoretically emphasizes the interrelatedness of all things. As such, Nietzsche's notion of the will to power suggests a paradigm of nature that comes close to the worldview of modern ecologists, a world in which nature is seen as a living process that is marked by continuous transformation.[10]

Again, in a general sense, Hallman is basically correct: along with deep ecologists, Nietzsche rejects traditional metaphysical categories such as substance and essence in favor of a dynamic, relational ontology. But one of the problems with Hallman's account is that he does not intuit that Nietzsche would have been skeptical about just how "subversive" the science of ecology actually is. Insofar as ecology frames natural phenomena in terms of a system of interrelated parts, it is arguably driven by the same objectivist perspective as the mechanistic or atomistic view it claims to subvert. In other words, when we embrace ecological scientific paradigms as "holistic" in contrast to the "dualistic" and "reductive" models in traditional empiricism or mechanistic views, we need to be careful that we are not falling into what we might call "ecologism," which would reduce beings to mere parts in an ecosystemic whole (Wilber, Zimmerman, and Esbjörn-Hargens refer to this as "subtle reductionism").[11] Moreover, Nietzsche would have rejected the Gaia hypothesis, the view that nature is a living thing or superorganism, which Hallman appears to impute to him.[12] At *GS* 109, Nietzsche explicitly says, "Let us beware of thinking that the world is a living being."[13] Ecologism risks losing the qualitative or hierarchical dimension in nature and can merely collapse humans to the level of just another animal species in an egalitarian "biotic soup."

But most important, Nietzsche does not call for the sheer dissolving of individuality. This is a common mistake among those who overemphasize the Dionysian dimension of his thought and miss his insistence on integrating the Dionysian and Apollonian spirits. Ralph Acampora correctly points out that Hallman overlooks Nietzsche's "aristocratic individualism," his reliance on hierarchy, and his insistence that life is inherently exploitative and appropriative, and thus "simplistically tames Nietzsche's feral philosophy."[14] Acampora thinks that although Hallman points out that Nietzsche expressed a sensitivity to nonhuman perspectives, he does not share the "organic axiology" of egalitarian environmental views like biocentrism and deep ecology. Nietzsche, Acampora writes, "prefers predatorial brutality to homely, herd-like traits,"[15] the wild over the domestic. I think he is correct on this point, since Hallman does not address the way that Nietzsche sees the need for a rank order of values and perspectives, but he himself fails to delve into "organic axiology" and takes much too literally the animal imagery of violence and predation that Nietzsche uses metaphorically to describe humans' inner struggles. Ultimately, Acampora is skeptical of using Nietzsche's positive vision of nature for environmental ethics.

The underlying problem here, in my view, is that Hallman and Acampora share a common prejudice: a one-sided view of hierarchy. Hallman tries to airbrush it out of Nietzsche's view, while Acampora thinks that it vitiates seeing Nietzsche as an environmental thinker. The assumption seems to be that an "environmentally correct" outlook has to espouse a nonhierarchical view of the human-nature relationship and that hierarchy inherently implies domination and exploitation. The problems with this assumption will become clearer in a moment.

Now let us turn to what I see as the more sophisticated accounts of Graham Parkes and Martin Drenthen. Drenthen claims that Nietzsche's thought poses a problem for environmental ethics because of a paradox in his view of interpretation: "Because each interpretation is necessarily contingent and restrictive, each environmental ethic that conceptualizes nature's intrinsic value relies on a conceptual and practical seizure of power over nature, similar to the one it wants to criticize."[16] Drenthen makes a useful distinction between "traditional environmental ethicists" and "postmodern environmental philosophers": the former tend to be naïve realists about nature who believe we have access to nature "in itself" and often base their conclusions on natural science, while the latter "argue that concepts such as nature and wilderness (signifying a realm opposed to culture) are social constructions that function within the cultural project of trying to control and understand reality."[17] Due to Nietzsche's view that "we can only know interpretations of nature and never nature as it is in itself,"[18] Drenthen thinks Nietzsche does not support any notion of natural value and that his view undercuts traditional environmental ethics. Since he (mistakenly, in my view) believes that for Nietzsche, "each value is man-made," Drenthen is led to conclude that "the awareness of the radical otherness of nature can lead to a new attitude of listening for nature and awareness of human finitude."[19] What we get, in other words, is something like Heidegger's *Gelassenheit*. But there are three problems here. First, all this gives us is the vague imperative to "let things be." As Drenthen concedes, "Sublime nature withdraws itself from us, it is inconceivable, and it provokes wonder and a feeling of awe. However, it does not allow us to identify its exact meaning or to construct a system of ethics that could justify our actions."[20] Second, as I argued in the previous chapter, Nietzsche does not believe that values are all man-made: all living things value insofar as they are composed of drives with goals leading to their preservation and enhancement. Third, Drenthen acknowledges that "the notion of 'the sublime' is itself yet another contingent attempt to identify the inconceivable in nature" and that for Nietzsche, "nature is also characterized by positive attributes such as creativity, greatness, forcefulness, independence, and necessity."[21] In effect, Drenthen presupposes that there is indeed a nature "in itself" beyond our interpretations. But since the sublime is just another construct, we have no grounds for embracing it and no justification for respecting its "alterity." To his credit, Drenthen realizes that "nature is still a key concept in [Nietzsche's] philosophy, contrary to post-modern theories. . . . These authors argue that Nietzsche's most important message is that we can only have a plurality of interpretations of nature. The problem with this approach is that the notion of 'nature' seems to have lost all meaning."[22] In other words, the constructivist path leads to nihilism. This is exactly the point I made in previous chapters in regard to Heidegger's later view of nature. It is not as though *Gelassenheit* and a poetic posture toward nature and ecological consciousness and so on are not important—they are. But I see *Gelassenheit* as something like a necessary but insufficient condition for a robust environmental ethic. The sublime can stop us in our tracks—and suspend scientific naturalism—but it cannot help us move forward.

Let us pause for a moment and probe this connection with poststructuralism because it ties together three of the main issues in this chapter: hierarchy, interpretation, and constructivism. The problem with the poststructuralist readings of Nietzsche to which Drenthen refers is that they cannot conceive of order or hierarchy in a positive, enabling, integrating manner. They tend to view schemata, interpretations, categorizations, and so on, as constructs laid over a kind of ineffable, chaotic flux that language and concepts can only conceal and distort; in this way, they are latter day versions of the Kantian sublime. But taken to its extreme, this implies a kind of dualism between human intentionality and interpretation, on the one hand, and unknowable nature "in itself," on the other. It reifies the void as something "on the other side" of the human horizon. And the major nondualistic, nonanthropocentric countermove to this approach that we have seen with Nietzsche and, at times, with Heidegger (recall his claim that the "earth" is not pure chaos, but is governed by some indecipherable law) is to concede some dimension and degree of interiority or intentionality to nonhuman beings.

One of the key issues here involves "social construction of nature" views.[23] Ted Toadvine sketches the problem with such views: "The constructivist view of nature holds that any 'access' to nature translates a certain function of discourse and ultimately our own self-reflections in the mirror of language, culture, and power."[24] For the constructivist, it is interpretation all the way down, and there is thus no view from nowhere, no standard by which to judge which interpretation is the most true; the consequence is skepticism. In a similar fashion, Steven Vogel perceptively illustrates how this problem developed in the romantics and *Lebensphilosophie* and runs through critical theory, Heidegger, and deep ecology:

> "[N]ature" and more generally that which is Other than the human or social takes on a positive sign, and contemporary science and technology are criticized . . . because they violate [nature's] otherness, its specificity as an ontological realm beyond the human and not fully graspable by it. . . . Not only is [this] the view defended by the great figures of the classical Frankfurt School . . . and by Marcuse as well, but its similarity to themes in the late Heidegger should also be clear. And of course versions of this sort of critique of science and technology are common in contemporary environmental discussions associated with "deep ecology" and related positions. But it faces . . . a series of deep and indeed ultimately fatal problems as a philosophical view, deriving fundamentally from *the difficulty it confronts in explaining how it can itself come to know the nature that it claims dominative worldviews fail to comprehend.* Its naturalism stands in conflict with its claims about the absolute otherness of nature, and the result is either incoherence or vacuity.²⁵

Or nihilism. This is why I believe "alterity theory"²⁶ is a dead end, and why we must pursue something like Evan Thompson's "strong continuity thesis" about the human-nature relationship. *Contra* Drenthen, Nietzsche cannot be classed as an alterity theorist for three reasons. First, Nietzsche believes that, in Jonas's phrase, "life can be known only by life." In the spirit of Jonas, Evan Thompson ambitiously yokes ideas from theoretical biology and phenomenology to explain how the continuity between life and mind can be secured. I think Nietzsche expresses a similar idea when he asks,

> Suppose nothing else were "given" as real except our world of desires and passions, and we could not get down, or up, to any other "reality" besides the reality of our drives. . . . Is it not permitted to make the experiment and to ask the question whether this "given" would not be sufficient for also understanding on the basis of this kind of thing the so-called mechanistic (or "material") world?²⁷

Human life can know nonhuman life because both are constituted by drives. "Will," he writes, "can only affect 'will'—and not 'matter' (not 'nerves,' for example)."²⁸ This—wills to power as the drive-life of all things—is what he refers to as "the world viewed from the inside . . . according to its 'intelligible' character."²⁹ This appears to be a difference from Heidegger. In *FCM*, Heidegger became bogged down in questions of methodological access to nonhumans rather than accept the evidence of our own experience as a clue to the subjective and evaluative aspects of living things. Second, unlike some poststructuralist thinkers (and Heidegger after *FCM*), Nietzsche is concerned to furnish a philosophical biology and a theory of evolution, whereas the latter tend to keep biology at arms' length and analyze it as a discourse whose concepts have no actual referent and say more about us than they do about life itself. Third, Nietzsche is a developmental thinker, like Hegel and Schelling before him. Though he rejects any sort of "end of history" or *telos* of evolution, there is no question that he views the will to power in general and the human psyche in particular in developmental terms. That there are discernible patterns in evolution need not rob it of mystery.

Now let us turn to Graham Parkes' account of Nietzsche's mature view of nature. Parkes also notes that Nietzsche's view of nature changed over time, yet his three-stage account is, I think, more nuanced than Hallman's:

> [F]rom [1] an early Romanticist view, through to [2] a sober, more rational understanding informed by modern science, to [3] a profound and comprehensive vision of humanity and the natural cosmos as dynamic and interpenetrating configurations of what he called "will to power." His final view, based on a reverence for the ultimately enigmatic nature of things, advocates a loyalty to the earth and a reverence for and affirmation of the "innocence" of natural phenomena in all their transience.[30]

Parkes' schema can not only help us track the development of Nietzsche's thoughts of life and nature but also show us why we should resist the claim that he is a vitalist (in the dismissive, pejorative sense of that term). In Nietzsche's middle period, he is more a scientific thinker trying to reduce religious, moral, and cultural phenomena to physiological causes and is generally skeptical of teleology and suspicious of metaphysics. It is this "debunking" spirit that invites, for example, a Brian Leiter to suppose that Nietzsche is a scientific naturalist. It is in Nietzsche's later period, in which the will to power rears its head as a metaphysical principle, that Nietzsche's position starts to appear more like vitalism. But it does not seem likely that the later Nietzsche would drop the "hard-nosed," scientific sensibility of his middle works completely; rather, I think that through his notion of the will to power as drives, he was attempting to work out a position that heeded the biology as closely as possible while leaving room for a minimal form of teleology along Richardson's and Thompson's lines.

Like Drenthen, Parkes sees the problem of constructivist versus realist views of nature and thinks that it grows out of the tension between the second and third stages. The second stage is driven by a skeptical and "debunking" spirit. As Parkes explains, "The salient feature of Nietzsche's coolly scientific phase of thinking about nature is his emphasis on the ways human conceptions of nature from epoch to epoch are conditioned by various kinds of fantasy projections, ranging from subjective caprice to impositions of humanly created regularities."[31] During this period, Nietzsche's task is to reduce what we might call the "ideological superstructures" of consciousness, religion, morality, and metaphysics to the "material bases" of drives; at times, he appears to verge on a kind of epiphenomenalism. However, he comes to believe that physics and positivism—and their concept of "objective nature"—are *also* interpretations of the world and that they too can be subjected to critique; this leads to a kind of constructivism. Yet Nietzsche also occasionally speaks of the "pure concept of 'nature'"[32] and a "newly redeemed nature."[33] At *BGE* 230, he writes the following:

> To translate the human back into nature; to become master over the many vain and fanatical interpretations and side-meanings that have so far been scribbled on that eternal ground-text *homo natura*. . . . to make it that the human being henceforth stand . . . before that *other nature* . . . deaf to the enticements of all the metaphysical bird-catchers who have been whistling to it for too long: "You are more! You are higher! You are of another origin!"[34]

And at section 188, he writes, "'Nature as it is, in all its extravagant and *indifferent* magnificence . . . appalls us, but is noble."[35] These seem to imply, contrary to the constructivist view, that there is a kind of nature "as it is" to which we might break through after all. As Parkes notes, "The tension between a view that understands fantasy projection as an ineluctable (if occasionally see-through-able) aspect of the human condition and one that allows for a seeing of the world of nature as it is in itself, apart from human projections on to it, persists to the time of Zarathustra."[36] This implies what Parkes terms a "twofold task": "to strip away the fantastic metaphysical interpretations of human origins that have obscured human nature, and to confront human beings with nature itself, similarly stripped of human projections."[37] Note that this twofold task corresponds to Nietzsche's project for values: a genealogy that traces extant values back to their source (negative)—our natural roots "appall" us when compared with our myths—and a revaluation that posits new values (positive)—nature "ennobles" us. Let us look more closely at Nietzsche's positive sense of nature.

The pivotal concept in Nietzsche's later, positive view of nature is the will to power. As I discussed in the previous chapter, there are three key aspects to it: its cosmic scope, its connection to interpretation, and its development along a continuum. In a sense, everything turns on whether will to power is interpreted anthropocentrically. As we saw, this is Heidegger's position: not that the will to power is merely a psychological principle, but that for Nietzsche, psychology becomes metaphysics through the projection of will onto all beings. But Parkes makes a strong case that Nietzsche's ultimate meaning for the concept is cosmic in scope. Discussing Nietzsche's "thought experiment of extrapolating from what we know immediately and intimately . . . to the rest of life and to 'the so-called mechanistic (or material) world,'" Parkes insists that

> this is by no means an instance of anthropocentrism, since Nietzsche has just desubstantialized the soul into a configuration of forces . . . shown the human "I" to be a fiction generated by the grammatical habit of positing a doer for every doing, and demonstrated "will" to be a complex function of forces issuing from a social structure of multiple "souls" deep within the body. Far from being the "will power" exerted by the human ego, the will of will to power is . . . a cosmic force.[38]

In other words, the problem of "access" to the nonhuman is canceled when the notion of a substantial self or soul is seen through; that is what permits Nietzsche's "extrapolation."

Will to power is an inherently interpretative force. This is usually taken in a strictly human sense, that given we are the only beings with language and self-consciousness, we are the only beings that interpret, that tell stories and create culture and give accounts of the way things are. However, Parkes suggests,

> If Nietzsche's suggestion that "all existence is essentially an interpreting existence" strikes us as strange, this is because we are so accustomed to the Cartesian dichotomy between the animate and inanimate (with only the human animate, *res cogitans*, being capable of interpreting). Less anthropocentric philosophies like Daoism and Mahayana Buddhism assume a continuum between

natural and human, with each particular on the continuum construing the world from its own perspective.[39]

Parkes acutely explains how the failure to see this interpretive aspect undermines previous efforts in seeing Nietzsche as an environmental thinker:

> Neither Hallman nor Acampora seems to appreciate this interpretive dimension of the will to power, Hallman being too focused on "the interrelated dimension of all things" and Acampora overemphasizing "exploitation." The latter rightly emphasizes the importance in Nietzsche of the order of rank and pathos of distance—but these are ideas that he applies to hierarchy among human beings and not to their putative superiority over natural beings.[40]

Nietzsche's perspective is best illustrated (if not most articulated) in Zarathustra. As Parkes explains, Zarathustra's ideal is to "let each particular thing generate its own horizons, arising and perishing just as it does. In terms of environmental ethics, to experience in this way allows one to appreciate the intrinsic value of the natural world absolutely."[41]

But what is the nature of this interpretive world projection? Interpretation and projection seem to imply a unidirectional imposition of the subject on the objective world. Parkes clarifies this by explaining just what it is the drives interpret: "There is some resistance there, something to 'push back' and set limits on how the world can be construed. . . . What pushes back . . . as our drives interpretively project a world, is will in the form of other drives—not only the drives of our fellow human beings, but also those that animate animals, plants, and other natural phenomena."[42] Perspectives, in other words, are always perspectives *on* other perspectives. A perspective is not a windowless monad but a finite clearing or opening within which the world manifests in a certain way, with some capacity for receptivity and response. At the human level, the higher or better interpretation will be the one that respects and incorporates as many "resistances" and "limits"—as many perspectives—as possible. So on the one hand, there is a "*Gelassenheit*" moment in Nietzsche's view that takes us beyond utilitarian or scientific perspectives on nature, but on the other hand, Nietzsche thinks it is incumbent on us to reconstruct nature as having a kind of structure, an evolving hierarchy of interrelated forces. All told, Parkes classifies Nietzsche as an "ecocentrist," the view that the main object of ethical concern should be biotic communities (wholes) rather than individual organisms (parts).

While Parkes gives the best account of Nietzsche's mature view of nature, I find three limitations in his view. First, he too seems to share the prejudice against hierarchy that hobbles Hallman and Acampora. Indeed, when he says that "the drives interpret at different levels of complexity,"[43] this implies that there is a spectrum of lower and higher and that, in the range of life forms, human beings are the most complex and that therefore when Nietzsche talks of rank order, he is not merely referring to human beings but also to the natural world. Indeed, at *BGE* 219, Nietzsche writes that the goal of a "high spirituality" is to "maintain the order or rank *in the world*, among things themselves—and *not only* among men."[44] Second, while Parkes acknowledges that Nietzsche embraces a notion of intrinsic value, since he eschews hierarchy, there is no way of ordering values along a vertical continuum; without the latter, value ceases to be a useful concept because we have no way of establishing priorities for action. Third, while his account of the importance of Nietzsche's psychology is unparalleled, he does not integrate this with Nietzsche's views on biology and evolution, which, as I attempted to show in the previous section, give us a firm footing for value theory. With all the voices in the debate now aired, I want to briefly outline my own position, circle back to Heidegger, and then explain how this nonreductive naturalism aligns with some other views in environmental ethics.

II. Hierarchical Biocentrism

It seems to me that Nietzsche's value theory strikes a balance between a strong anthropocentrism that would deny intrinsic value to nonhuman entities, on the one hand, and a nonanthropocentric, biocentric view that recognizes the intrinsic value of nonhumans yet equates it with that of humans. While Nietzsche's interests may not have been aligned with those of today's environmental thinkers, I think his view provides a check against the misanthropy that can sometimes creep into nonanthropocentric discourse.[45] In mainstream environmental ethics, his position closely resembles Paul Taylor's biocentrism. While Taylor is committed to a bioegalitarianism that Nietzsche would reject, he holds that humans are able to empathetically enter into and grasp the perspectives and interests of nonhuman species and concedes that in practice, we need a "weighing mechanism" to handle interspecific conflicts of interest. Taylor enumerates "priority principles" for mediating such conflicts that justify overriding nonhuman interests for those human interests that are based on our particular "teleological perspective" as rational beings and holds that we should not be willing to give up said interests for the sake of the nonhuman; as such, he seems to embrace an ontological hierarchy by basing interests on something like an Aristotelian teleological view of life (albeit an evolutionary one).[46] Though we want to expand our traditional ethics beyond the human purview, that does not mean we need to demonize intrinsic human capacities and endeavors such as reason, science, and technology; though the latter can certainly overstep their bounds and cause great environmental havoc, they need not be seen as the enemy. Nietzsche offers us an account of our kinship with animals and other natural beings, yet his attention to the problem of nihilism keeps his eye on how human flourishing, though in part dependent on its natural foundation, is a distinct sphere with its own value.

According to Nietzsche's value theory, all beings have moral standing on account of being constellations of drives with certain ends. Humans, to speak crudely, have more "moving parts" than amoebas—they are composed of and integrate a greater variety of drives, enjoy a richer and more intense experience of the world, and are "more powerful" in the sense that they have a greater range of possibilities, a greater capacity to create, and embrace more of reality. For Nietzsche, the strength or power of an entity depends on the number of drives it composes and integrates. While reason, too, is another dimension of drive life, not opposed to it—Kaufmann casts Nietzsche's ideal as the "passionate man who masters his passions"[47]—it is unique in its critical power to repress and negate pre-rational drives. That is why reason, for Nietzsche, is such a threat to life. But his idea seems to be that, when no longer opposed to the passions or bent on control of external beings, reason has a creative, affirmative power and is the means by which we can escape the narcissism of petty drives and, as we saw Taylor put it above, project ourselves into, resonate with, and in some sense take responsibility for, the perspectives of other beings. Kaufmann, quoting Nietzsche, corrects the popular view of his thought as "irrationalist": "Rationality 'distinguishes the higher from the lower men.' Nor is this a casual point in Nietzsche's writings. The identification of the hatred of reason with the bad intellectual conscience can be found everywhere in his books and notes; irrationality is ever a weakness in his eyes; and rationality, a sign of power."[48] As such, I think that on Nietzsche's account we can safely say that humans are more valuable than, say, amoebas. However, this is not to say that amoebas have no value. Insofar as they are driven beings, they realize some value; this may be so slight as to be vastly less than other forms of life, especially humans, but it must be acknowledged in order to preserve the ontological continuity between humans and the rest of life.

But I think there is another kind of value implied in Nietzsche's account. If my account of Nietzsche's view of hierarchy is correct, then "positive" hierarchies are those in which the higher does not dominate and exploit the lower, but incorporates it in such a way as to open it up for possibilities it could not realize on its own. Successful, powerful life forms are those in which there is, as it were, a "power-sharing agreement," a mutual dependence between its higher and lower drives. For instance, the behavior of a cell unlocks potentials in the molecules that compose it that the latter could never have realized on their own. If we expand this logic beyond the organism and into the ecological domain—that between organisms and species—then what we get is something like a "power chain": "lower" organisms have symbiotic relationships with higher ones insofar as they support the higher and are freed for new possibilities by the higher. Erazim Kohak cogently explains this idea through his understanding of extrinsic value. He argues that the recognition of a being's extrinsic value should not be interpreted as eviscerating its being. The extrinsically valuable aspect of a being *really is* valuable. Kohak thinks that the "crux of the problem . . . is the recognition of what I would describe as the *intrinsic value of utility*. There are times when being treated *as a means* is precisely the respect due to a being as an end. . . . I would speak also of a strong sense that things deserve, 'have a right' to be used."[49] Indeed, it is our attempts to "keep our hands clean" and refrain from "contaminating" nature by leaving it to its own designs that constitutes abuse. As Kohak eloquently puts it, "it is the callousness of sentimentality rather than the putative rapacity of reason" that, paradoxically enough, leads us to sever our kinship with nature."[50] He uses two examples to illustrate this point: the culling of horse herds in the Grand Canyon and the thinning of trees in a forest. In both cases, a lack of human intervention, born by an excess of reverence and appreciation for the beauty of nature, led to long-term environmental devastation. He concludes that "for better or worse, we have become part of the balance of nature and can no longer simply withdraw."[51] In short, though lower drives or organisms or species may have less intrinsic value, they have another kind of value that derives from the supportive function they play for higher ones. The key is that all beings have both kinds of value, only in varying degrees.

I think that this sketch is basically congruent with two other positions in environmental ethics, both of which recognize the interiority and value of nonhumans and embrace a hierarchical cosmology: Michael Zimmerman's and Sean Esbjörn-Hargens' "integral ecology" and David Ray Griffin's "process ecology." Integral ecology recognizes both "intrinsic" and "extrinsic" forms of value. Roughly put, the former refers to the value a thing has in and for itself, while the latter refers to the value a thing has for others. Zimmerman and Esbjörn-Hargens explain these forms in terms of the biosphere and, using Teilhard de Chardin's phrase, the noosphere:

> In terms of extrinsic value, the biosphere is primary because it is more fundamental. If we were to destroy the biosphere, we would also destroy the noosphere. Thus, the biosphere is primary, and this means that the biosphere is part of us. The noosphere is not part of the biosphere. If it were, the biosphere could not exist without us. But the opposite is true. Many environmentalists intuit this, but they confuse what is most fundamental (Gaia/biosphere), or what has the most extrinsic value, with what is most significant (humans/noosphere), or what has the most intrinsic value.[52]

So the higher is more significant or meaningful—in Nietzsche's terms, more powerful—while the lower is more fundamental and supportive. The mistake of some environmentalists is to think that the emergence of the human—or agriculture, or industry, or capitalism, and so on—was some sort of fall from grace that disrupted a "natural harmony" and that we must dispense with any hierarchical view that holds humans to be special on account of their soul or intellect and reduce them to mere animals. The paradox of this is that, by attempting to "spread the wealth" of intrinsic value around equally, the concept loses meaning, because we are actually confusing two different kinds of value.

Zimmerman and Esbjörn-Hargens describe a similar problem that they dub the "fundamental paradox of environmentalism." The paradox is that many environmentalists value nature but endorse a value-free view of nature; in other words, their account of nature cannot account for their own accounting and evaluative activities:

> Environmentalists value the natural world but typically subscribe to a conception of nature that either excludes value (subjective and intersubjective perspectives) or regards it as a conventional fiction for enhancing human survival. . . . Environmentalists often speak of nature as a complex dynamic system in which humans, like other animals and plants, are merely strands in a cosmic web that lacks any hierarchy or direction. Yet, if humans are merely strands in a complex state of affairs—the *is*—they

are in no way capable of calling for alternative actions based on moral obligation—the *ought*.[53]

Both horns of this paradox are nihilistic because they assume a nonnormative conception of nature. As I have stressed above, Nietzsche evades this problem because his view of nature allows some measure of subjectivity and valuation to nonhumans. Given this, it is safe to say that Nietzsche would have found some strains of today's environmental thinking and activism to be nihilistic. Zimmerman aptly notes that Nietzsche would likely have concluded "that environmentalists are in many cases ascetics who fail to posit an adequate goal for humanity."[54]

David Ray Griffin also distinguishes between intrinsic and extrinsic value. Griffin explains that a key aspect of extrinsic value is "ecological" value: "A most important dimension of the extrinsic value of something it its ecological value, meaning its value for sustaining the cycle of life. Its value as food for other beings would be part of this ecological value, but so would many other functions, such as the function of worms in aerating the soil and that of certain soil bacteria in nitrogen fixation."[55] Griffin exploits the distinction between intrinsic and ecological value in order to solve a problem in Arne Naess's deep ecology. Naess, along with a great many deep ecologists and environmental thinkers, held that all living things have equal intrinsic value yet, as is widely known, he conceded that it is practically impossible to act in accord with this principle. Griffin thinks this problem is solved when we see that there is an "inverse correlation" between intrinsic and ecological value:

> Those species whose (individual) members have the least intrinsic value, such as bacteria, worms, trees, and the plankton, have the greatest ecological value; without them, the whole ecosystem would collapse. By contrast, those species whose members have the greatest intrinsic value (meaning the richest experience and thereby the most value for themselves), such as whales, dolphins, and primates, have the least ecological value.[56]

Griffin submits that the basic intuition of biospheric or egalitarian deep ecology—namely, that all living things have equal value—is correct so long as we distinguish between "intrinsic" value (the value a thing has for itself) and "inherent" value (the total value of a thing, i.e., its intrinsic and ecological values); inherently, then, all living things are of equal value, but intrinsically or ecologically, they are not. Without this distinction, the hierarchical dimension of nature is lost, and we are plunged into confusion about how the bioegalitarian ideal and our "bias" for humans and higher animals can coexist.

To be sure, there are still difficult issues raised by the account I have advanced. It is one thing to claim that "x values"; it is another thing to claim that "x is valuable." Tom, Dick, and Harry might value a, b, and c, yet the mere fact of their valuing does not tell us which interest should by given priority. Just so, a human, an owl, and a chipmunk might value a, b, and c, but that does not by itself tell us which interest ought to be given priority. A number of thinkers who, like Nietzsche, hold that nonhumans value or at least have interests worthy of consideration, have offered criteria and desiderata for adjudicating such conflicts.[57] A comparative analysis of Nietzsche's view with these frameworks may yield concrete prescriptions regarding environmental problems. My main purpose here, however, has been to critique aspects of Heidegger's later view of nature—notably, its neglect of evolution and value theory—and present aspects of his earlier thought that are congruent with Nietzsche's view as a more viable view of nature on which an ethic of the environment might be built. While these parallels with Nietzsche's implicit environmental ethics are general and only roughly sketched here, and while his own position is often not adequately or consistently articulated in his own works, I think my outline of his mature view of nature and value theory offers a departure from the alterity theory that has dominated continental environmental thought.

In closing, let me underscore a few key conclusions. I began by insisting that before we can determine our duties to nature, we must determine what we mean by nature. We saw that Heidegger's thought has been taken by many as a solution to the ecological crisis because of his critiques of the anthropocentric and metaphysical foundations of modern science and technology, forces often held responsible for environmental destruction. While his criticism of the dominant view of nature in modernity is compelling, and though some of his ideas are promising for environmental ethics, his alternative is not sufficient because it does not incorporate evolutionary theory, it seems to hold that nature is not intelligible, it fails to grapple with the gap between animals and humans, and it eschews the notion of value. Nietzsche also perceived the nihilistic implications of the modern view of nature; indeed, as Kaufmann puts it, just as Kant was awakened by Hume, Nietzsche was awakened from his dogmatic slumber by Darwin. His account of nature can supplement Heidegger's by incorporating evolutionary theory yet retaining the "depth" dimension of traditional views of nature such as the great chain of being and insisting that valuation is intrinsic to all life. While they never formulated a "system of values" or an explicit environmental ethic, their original views of life, value, and nature—conceived as responses to the problem of nihilism—present an orienting vision for environmental ethics.

Notes

Introduction

1. E. A. Burtt, *The Metaphysical Foudations of Modern Science* (New York: Doubleday Anchor Books, 1954), 17–18.

2. Keith Campbell, "Naturalism," in *Encyclopedia of Philosophy*, ed. Donald Borchert, Volume 6, 492–95, 492.

3. Mario De Caro and David Macarthur, eds., *Naturalism in Question* (Cambridge: Harvard University Press, 2004), 10.

4. Ibid., 4.

5. See, for instance, William Lane Craig and J. P. Moreland, eds., *Naturalism: A Critical Analysis* (New York: Routledge, 2000); Owen Flanagan, *The Really Hard Problem: Meaning in a Material World* (Cambridge: MIT Press, 2007); Brian Lightboy, *The Problem of Naturalism: Analytic Perspectives, Continental Virtues* (New York: Lexington Books, 2013); John R. Shook and Paul Kurtz, eds., *The Future of Naturalism* (Amherst: Humanity Books, 2009); Evan Thompson, *Mind in Life: Biology, Phenomenology, and the Sciences of Mind* (Cambridge: Harvard University Press, 2007); and the anthology edited by DeCaro and Macarther cited above.

6. Alfred North Whitehead, *Science and the Modern World* (New York: The Free Press, 1967), 80.

7. Cited by Burtt, 46.

8. Of course, one might object that this does not follow, since our duties to humans may entail restrictions on how we treat the environment. I would reply that while this may be the case, the point is that it only *may* be the case: in the absence of any human interest, there need be no regard for how our actions affect, e.g., animals, species, habitats, and so on.

9. Ted Benton, "Naturalism in Social Science," in *Routledge Encyclopedia of Philosophy*, ed. E. Craig (London: Routledge, 1998), retrieved August 8, 2010, from http://www.rep.routledge.com/article/RO11.

10. Though, to be clear, he eventually abandons transcendental phenomenology in his later work, he retains its antinaturalist posture.

Chapter 1

1. Michel Haar. *The Song of the Earth: Heidegger and the Grounds of the History of Being*, trans. Reginald Lilly (Bloomington: Indiana University Press, 1993), 10.

2. For an overview of the role of primal Christianity in Heidegger's early thought, see John van Buren, *The Young Heidegger* (Bloomington: Indiana University Press, 1994), chapter 8. For his lecture courses on Aristotle, given in the early 1920s, see Martin Heidegger, "Phenomenological Interpretations in Connection with Aristotle: an Indication of the Hermeneutic Situation," in *Supplements: From the Earliest Essays to* Being and Time *and Beyond*, ed. John van Buren (Albany: SUNY Press, 2002), 111–47; *Phenomenological Interpretations of Aristotle*, trans. Richard Rojcewicz (Bloomington: Indiana University Press, 2001) (*GA* 61); "Being-There and Being-True According to Aristotle," trans. Brian Bowles, in *Becoming Heidegger: On the Trail of His Early Occasional Writings, 1910–1927*, ed. Theodore Kisiel and Thomas Sheehan (Evanston: Northwestern University Press, 2007), 214–38. For Heidegger's early interpretations of Aristotle as precursors to *Being and Time*, see *The Young Heidegger*, chapter 10; Franco Volpi, "*Being and Time*: A 'Translation' of the *Nicomachean Ethics*?" in *Reading Heidegger from the Start: Essays in His Earliest Thought*, ed. Theodore Kisiel and John van Buren (Albany: SUNY Press, 1994), 195–213; Theodore Kisiel, *The Genesis of Heidegger's Being and Time* (Berkeley: University of California Press, 1993); Thomas Sheehan, "On the Way to *Ereignis*: Heidegger's Interpretation of Physis," in *Continental Philosophy in America: Prize Essays, Volume I*, ed. John Sallis (Pittsburgh: Duquense University Press), 131–64.

3. I think one reason Heidegger's (and Nietzsche's) work has sparked such interest among Buddhist philosophers is that he rejects the category of substance and recognizes the significance of the nothing—but this is still a kind of metaphysics, if we understand that broadly to be an account of the being of beings or nature of reality. See Graham Parkes, ed., *Heidegger and Asian Thought* (Honolulu: University of Hawaii Press, 1987).

4. Broadly construed as starting with Heraclitus, to some extent moving through Leibniz and Spinoza, the German idealists, Nietzsche, and through the pragmatists and Whitehead to David Ray Griffin today.

5. More specifically, around the 1930s Heidegger comes to see science and philosophy as fundamentally opposed. I address the evolution of Heidegger's view of the relationship of science and philosophy throughout. See Trish Glazebrook, *Heidegger's Philosophy of Science* (New York: Fordham University Press, 2000).

6. See John Caputo, *Against Ethics* (Bloomington: Indiana University Press, 1993); Michael Lewis, *Heidegger and the Place of Ethics* (New York: Continuum, 2005).

7. Hans Jonas, *The Phenomenon of Life* (New York: Harper and Row, 1966), 253. Jonas is one of the pivotal figures in the neglected tradition of "bio-phenomenology" that I mention below and advert to in later chapters.

8. Karl Löwith, *Nature, History, and Existentialism*, ed. Arnold Levison (Evanston: Northwestern University Press, 1966), 18.

9. Ibid., 29.

10. Zimmerman, who inaugurated the Heidegger-deep ecology dialogue and initially believed the two were compatible, has grown more skeptical of the connection and critical of Heidegger's position over the years. See his "Toward a Heideggerian Ethos for Radical Environmentalism," *Environmental Ethics* 6, no. 2 (1983); "Rethinking the Heidegger—Deep Ecology Relationship," *Environmental Ethics* 15 (1993): 95–224; and chapter 3 of *Contesting Earth's Future*. Also see Stephen Avery, "The Misbegotten Child of Deep Ecology," *Environmental Values* 13, no. 1 (February 2004): 31–50; Bill Devall and George Sessions, *Deep Ecology: Living as if Nature Mattered* (Salt Lake City: Peregrine Smith Books, 1985).

11. Fox, for instance, attempted to pin the movement on Naess' spiritually tilted norm of "self-realization," while Andrew McLaughlin argues that this norm is based on religious and spiritual, i.e., nonrational, grounds, and should thus be bracketed, leaving only the widely accepted eight-point platform to serve as the "heart of Deep Ecology." See George Sessions, "Deep Ecology: Introduction," in Zimmerman, *Environmental Philosophy*, 1998, 173. Naess tries to negotiate this debate by claiming that "Ecosophies are not religions in the classical sense. They are better characterized as general philosophies, in the sense of total views, inspired in part by the science of ecology." See Arne Naess, "The Deep Ecological Movement," in Zimmerman, *Environmental Philosophy*, 207. We might say that deep ecologists are united around a set of propositions supported by ecological science, yet frame this set of propositions in terms of a particular religious or spiritual worldview or set of worldviews.

12. Martin Heidegger, *Introduction to Metaphysics*, trans. Ralph Mannheim (New York: Yale University Press, 1959), 13 (*GA* 40).

13. Michael Zimmerman, "Toward a Heideggerian Ethos for Radical Environmentalism," *Environmental Ethics* 6, no. 2 (1983); "Marx and Heidegger on the Technological Domination of Nature," *Philosophy Today* 23 (Summer 1979): 99–112; "Implications of Heidegger's Thought for Deep Ecology," in *The Modern Schoolman* 64 (November 1986): 19–43. See also Laura Westra, "Let It Be: Heidegger and Future Generations," *Environmental Ethics* 7, no. 4 (1985): 341–50, who makes a similar argument that Heidegger's nonhierarchical view of nature and of humans as "freeing" things to exist in their own way offers a sounder basis than utilitarian or rights-based ethics for protecting future generations of humans and nonhumans.

14. Devall and Sessions, *Deep Ecology*, 98–100.

15. Ibid., 74.

16. See Sheehan essays and van Buren chapter in note 2 above. I discuss Heidegger's early engagement with Aristotle in detail in chapter 2.

17. *Supplements*, 10–11.

18. Tracy Colony, "Dwelling in the Biosphere? Heidegger's Critique of Humanism and Its Relevance for Ecological Thought," *International Studies in Philosophy* 31, no. 1 (1999): 37–45.

19. Nancy Holland, "Rethinking Ecology in the Western Philosophical Tradition: Heidegger and/on Aristotle," *Continental Philosophy Review* 32 (1999): 409–20.

20. Charles Taylor, "Heidegger, Language, and Ecology," in *Heidegger: A Critical Reader*, ed. Hubert Dreyfus (Cambridge: Blackwell, 1992).

21. See the essays in Ladelle McWhorter, ed. *Heidegger and the Earth* (Kirksville: Jefferson University Press, 1992).

22. Ibid., 175.

23. See *Contesting Earth's Future* and "Rethinking the Heidegger—Deep Ecology Relationship," *Environmental Ethics* 15 (1993): 95–224; Foltz and DeLuca cited above; Leslie Paul Thiele, "Nature and Freedom: A Heideggerian Critique of Biocentric and Sociocentric Environmentalism," *Environmental Ethics* 17, no. 2 (Summer 1995): 171–90; Stephen Avery, "The Misbegotten Child of Deep Ecology," *Environmental Values* 13, no. 1 (February 2004): 31–50; Daniel Dombrowski, "Heidegger's Anti-Anthropocentrism," *Between the Species* 10, nos. 1–2 (1994): 26–38.

24. Foltz, 176.

25. Ibid., 121.

26. Charles S. Brown and Ted Toadvine, "Eco-Phenomenology: an Introduction," in *Eco-Phenomenology: Back to the Earth Itself* (Albany: SUNY Press, 2003).

27. Erazim Kohak, *The Embers and the Stars* (Chicago: University of Chicago Press, 1984), xii.

28. John Llewelyn, *The Middle Voice of Ecological Conscience: A Chiasmic Reading of Responsibility in the Neighborhood of Levinas, Heidegger, and Others* (New York: St. Martin's, 1991); David Abram, *The Spell of the Sensuous* (New York: Pantheon Books, 1996). See also Frank Schalow, *The Incarnality of Being* (Albany: SUNY Press, 2006), for a study of Heidegger's explorations of the body, the human-animal relationship, and the alterity of the earth which claims that embodiment is the pivotal concept for understanding and navigating our ecological situation.

29. *Middle Voice*, 254–55.

30. David Farrell Krell, *Daimon Life: Heidegger and Life-Philosophy* (Bloomington: Indiana University Press, 1992); Brett Buchanan, *Onto-Ethologies: The Animal Environments of Uexküll, Heidegger, Merleau-Ponty, and Deleuze* (Albany: SUNY Press, 2008). I return to these important texts in later chapters.

31. Krell, xi. I will examine Krell's work in detail in later chapters to support the claim that Heidegger's thoroughgoing nonevolutionary understanding of life is a serious obstacle to a viable philosophy of nature.

32. Foltz, 22.

33. Parvis Emad, "Heidegger's Value-Criticism and Its bearing on the Phenomenology of Values," *Research in Phenomenology* 7 (1977):, 194.

34. Thomas Nenon, "Values, Reasons for Actions and Reflexivity," in *Phenomenology of Values and Valueing*," ed. James Hart and Lester Embree (Boston: Kluwer Academic Publishers, 1997, 117–37), 129.

35. Martin Heidegger, *Being and Time*, trans. John Macquarrie and Edward Robinson (New York: Harper and Row, 1962). 190–91 (*GA 2*).

36. "Heidegger's Phenomenology," 74.

37. Keith Campbell, "Naturalism," in *Encyclopedia of Philosophy*, ed. Donald Borchert, volume 6, 492.
38. Mario De Caro and David Macarthur, eds., *Naturalism in Question* (Cambridge: Harvard University Press, 2004), 10.
39. Ibid., 3.
40. Ibid., 4.
41. However, there are many thinkers in both the analytic and continental traditions who see major problems with scientific or "reductive" naturalism and have attempted to work out nonreductive versions of naturalism.
42. Though I will at times refer to phenomenological or nonreductive naturalism, which are importantly different, context should make clear which sense I intend.
43. Ibid., 493.
44. *Eco-Phenomenology*, xi.
45. This is not to claim that phenomenology only deals with "first-person experiences," since this would already be making a metaphysical claim about subjectivity that the phenomenological reduction is designed to suspend.
46. Ibid., 39.
47. Ibid., 47.
48. Ibid., xii.
49. *Being and Time*, 132–33, 190, 258.
50. *Basic Writings*, ed. David Farrell Krell (New York: Harper and Row, 1977), 251 (*GA* 9).
51. Neil Evernden, *The Natural Alien* (Toronto: University of Toronto, 1993); Marjorie Grene, *Philosophical Approaches to Biology* (New York: Basic Books, 1968); Hans Jonas, *The Phenomenon of Life* (New York: Harper and Row, 1966); Karl Loewith, *Nature, History, and Existentialism*, ed. Arnold Levison (Evanston: Northwestern University Press, 1966); Max Scheler, *Man's Place in Nature*, trans. Hans Meyerhoff (New York: The Noonday Press, 1961).
52. Ibid., xii.
53. For Heidegger's treatments of the *Umwelt*, see *Being and Time*, 94–107, and *Fundamental Concepts of Metaphysics*, trans. William McNeill (Bloomington: Indiana University Press, 2001) (*GA* 29/30), part 2, chapters 2–5, and Buchanan, chapters 2 and 3.

Chapter 2

1. Didier Franck, "Being and the Living," in *Who Comes After the Subject?*, ed. Cadava, Connor, and Nancy (New York: Routledge, 1991), 146.
2. Jacques Derrida, *Of Spirit: Heidegger and the Question*, trans. Geoffrey Bennington and Rachel Bowlby (Chicago: University of Chicago Press, 1989), 105.
3. Giorgio Agamben, *The Open: Man and Animal*, trans. Kevin Attell (Stanford: Stanford University Press, 2004), 39.

4. Krell, 17.
5. *FCM*, 188–89.
6. Jakob von Uexküll, *A Foray into the Worlds of Animals and Humans*, trans. Joseph D. O'Neil (Minneapolis: University of Minnesota Press, 2010), 45.
7. Ibid., 135.
8. Ibid., 52.
9. Ibid., 189.
10. Glazebrook, 5–6. It is not an accident that Glazebrook's book contains not one entry in the index for the terms "biology" or "life." As she shows, Heidegger's conception of science was all but unilaterally determined by modern mathematical physics and the Kantian epistemological-metaphysical complex that grew around it.
11. Cited by Dorion Sagan in Uexküll, 4.
12. Quoted in Krell, 17.
13. I concur with Stuart Elden's claim that, with regard to Heidegger's views of animal life, we should "pay attention not merely to *The Fundamental Concepts of Metaphysics*, but to the much wider range of studies of animals that he undertakes throughout his career. Looking at these shows that Heidegger's work on the question of the animal can thus be understood as central to a number of his most important claims. The analysis of the animal's way of existing sheds valuable light on human Dasein, and the question of animality is linked to the central questions of language, politics and calculation." Stuart Elden, "Heidegger's Animals," *Continental Philosophy Review* 39 (2006):, 280.
14. Krell, 35.
15. Martin Heidegger, *Ontology—The Hermeneutics of Facticity*, trans. John van Buren (Bloomington: Indiana University Press, 1999) (*GA* 63); *Phenomenological Interpretations of Aristotle*, trans. Richard Rojcewicz (Bloomington: Indiana University Press, 2001) (*GA* 61); *Supplements: From the Earliest Essays to* Being and Time *and Beyond*, ed. John van Buren (Albany: SUNY Press, 2002); *Basic Concepts of Ancient Philosophy*, trans. Richard Rojcewicz (Bloomington: Indiana University Press, 2008) (*GA* 18); *Basic Concepts of Aristotelian Philosophy*, trans. Robert D. Metcalf and Mark B. Tanzer (Bloomington: Indiana University Press, 2009) (*GA* 18).
16. *Phenomenological Interpretations of Aristotle*, 83.
17. Supplements, 143. I discuss his fragmented sketch of this ontology of life below.
18. Krell, 39.
19. Jonas, 86.
20. *Basic Concepts of Ancient Philosophy*, 266–67.
21. *Phenomenological Interpretations of Aristotle*, 145.
22. Ibid.
23. Krell, 49.
24. *Basic Concepts of Ancient Philosophy*, 153.
25. Ibid., 156.
26. Ibid., 228.
27. Ibid., 154.

28. Buchanan, 23–24.

29. Ibid., 28.

30. Ibid., 35.

31. For an excellent article critiquing Heidegger for falling prey to the "prejudice of the lone animal," that is, ignoring the intrinsically social dimension of animals, that draws on research in neuroscience and cognitive ethology, see David Morris, "Animals and Humans, Thinking and Nature," *Phenomenology and the Cognitive Sciences* 4 (2005): 49–72. For a defense of animal minds on Heideggerian grounds, see Simon James, "Phenomenology and the Problem of Animal Minds," *Environmental Values* 18 (2009): 33–49.

32. *Basic Concepts of Aristotelian Philosophy*, 39.

33. Ibid., 33.

34. Ibid., 39.

35. For a discussion of the human-animal distinction along these lines, see Elden, 280.

36. *Basic Concepts of Aristotelian Philosophy*, 37.

37. Cited by Buchanan, 92–93. By "biology" Heidegger seems to be referring to anti-Darwinist trends led by Uexküll.

38. Buchanan, 91.

39. Martin Heidegger, *Logic: The Question of Truth*, trans. Thomas Sheehan (Bloomington: Indiana University Press, 2010), 215–66 (*GA 21*), my emphasis. Heidegger's dismissal of Uexküll here is based on a facile distinction between "philosophy" and "biological research"; as Buchanan demonstrates, Uexküll's understanding of animal *Umwelts* was based on a creative application of Kant's first critique. See Buchanan, chapter 1. Incidentally, Heidegger's dismissal of Darwinism may very well be a function of Uexküll's uncharitable interpretation of Darwin.

40. Martin Heidegger, "Vom Wesen der Wahreit," unpublished transcript of Heidegger's lecture in December 1924 (translated by John van Buren); the transcript can be found in the Helene Weiss Archive at Stanford University.

41. In *Being and Time*, "*Umwelt*" is only discussed with reference to humans and refers to the immediate surroundings of prereflective human Dasein in relation to its present dealings, whereas "world" refers to the wider totality of significance of which the *Umwelt* is only a part. In *FCM*, Umwelt is said to belong to animals, as the closed circuit of their drives in relation to their milieu, but this is different from world, which humans are open to because they have speech, interpretation, understanding, and temporality; however, Heidegger is cagey about whether animals do or do not have world, which perhaps reflects his lingering attachment to Aristotle's *scala natura* view.

Chapter 3

1. *Being and Time*, 22.
2. Ibid., 39.

3. Ibid., 63–64.
4. Ibid., 25.
5. Ibid., 67.
6. Ibid.
7. As I explain in the following chapters, in and after *Being and Time*, this quasi-Kantian, transcendental approach comes to dominate Heidegger's thought in another guise: the history of being. This has crucial consequences for his philosophy of nature.
8. Ibid., 234.
9. Hubert Dreyfus, *Being-in-the-World: A Commentary of Heidegger's Being and Time*, division 1 (Cambridge: MIT Press, 1991), 35.
10. *Being and Time*, 276.
11. Ibid., 375.
12. Ibid., 377.
13. Taminiaux, 47.
14. See John van Buren, *The Young Heidegger: Rumor of the Hidden King* (Bloomington: Indiana University Press, 1994), 122–29 and 352–57, and Michael E. Zimmerman, *Heidegger's Confrontation with Modernity* (Bloomington: Indiana University Press, 1990), 18–27.
15. Supplements, 37.
16. Ibid., 357.
17. *Being and Time*, 91.
18. Ibid., 92.
19. Ibid., 440–41.
20. Ibid., 87.
21. Ibid., 93.
22. Ibid., 91.
23. Ibid., 94.
24. Haar, 10.
25. *Being and Time*, 83.
26. Ibid., 93.
27. Ibid., 92.
28. Ibid., 93.
29. Ibid., 94.
30. Ibid., 97. This point is important, since it correlates with the notion of "falling" discussed later in the text. The image calls to mind the notion that, at a fundamental level, we are always plummeting toward the world and that we attempt to prop ourselves up by latching on and clinging to what is around us in order to break our fall. The picture of nature that emerges will depend on the way in which we "fix" the world—what we hold on to and how we hold on. The sense of nature is, in this way, a product of our response to nihilism.
31. Ibid., 99.
32. Ibid., 143.

33. Contemporary examples are E. O. Wilson's sociobiology or Daniel Dennett's presentation of Darwinism as a "universal acid" that eats through all disciplinary/regional barriers.
34. Buchanan, 40.
35. *Being and Time*, 30.
36. Ibid., 75.
37. For further details, see Buchanan, chapter 2, pp. 39–55.
38. Ibid., 84–85.
39. Krell, 37.
40. Ibid., 38.
41. Ibid., 81.
42. Ibid.
43. Krell, 50.
44. Ibid., 72.
45. Bruce Foltz, *Inhabiting the Earth: Heidegger, Environmental Ethics, and the Metaphysics of Nature* (New Jersey: Humanities Press, 1995), 31–32.
46. *Being and Time*, 100, my emphasis.
47. *Being and Time*, 101.
48. Foltz, 25.
49. Ibid., 22.
50. *Being and Time*, 481.
51. Foltz, 31. Foltz is referring primarily here to Jonas and Löwith.
52. Quoted by Foltz, 13. I will revisit the underlying issue of Heidegger's position regarding scientific realism in my discussion of Dreyfus's interpretation below.
53. *Being and Time*, 100.
54. Ibid., 94.
55. Foltz, 13.
56. Ibid., 14.
57. Dreyfus, 111.
58. Ibid.
59. Ibid., 109.
60. Ibid., 120.
61. Ibid., 111.
62. *Being and Time*, 101.
63. Ibid.
64. Ibid.
65. Ibid., 102.
66. Dreyfus, 252.
67. Ibid., 255.
68. Quoted by Dreyfus, 255.
69. Ibid.
70. Foltz, 42.
71. Ibid., 256.

72. Ibid., 264.
73. Ibid., 34.
74. Dreyfus, 110.
75. Hubert Dreyfus, "How Heidegger Defends the Possibility of a Correspondence Theory of Truth with respect to the Entities of Natural Science," posted on the author's website.
76. Ibid., 258.
77. Graham Parkes, "Thoughts on the Way," in *Heidegger and Asian Thought*, ed. Graham Parkes (Honolulu: University of Hawaii Press, 1987), 142 n26.
78. *Contesting Earth's Future*, 115, my emphasis.
79. Ibid., 115.
80. Quoted by Dreyfus, 256.
81. Quoted by Dreyfus, 259.
82. *Being and Time*, 441.
83. Dreyfus, 259.
84. Cf. David Ray Griffin, "Whitehead's Deeply Ecological Worldview," in *Worldviews and Ecology*, ed. M. E. Tucker and J. Grim (Maryknoll, NY: Orbis Books, 1994).
85. Ibid., 259, 256.

Chapter 4

1. David Farrell Krell, *Daimon Life: Heidegger and Life-Philosophy* (Bloomington: Indiana University Press, 1992). I will draw deeply on Krell's monograph throughout the chapter.
2. Frank Schalow, "Essence and Ape: Heidegger and the Question of Evolutionary Theory," *American Catholic Philosophical Quarterly* 82, no. 3 (2008): 445.
3. Iain Thompson, "Ontology and Ethics at the Intersection of Phenomenology and Environmental Philosophy," *Inquiry* 47 (2004): 381.
4. Ibid., 384.
5. Ibid., 385.
6. Note how this distinction roughly tracks the split between the two phenomenological traditions I discussed in chapter 1, "transcendental" and "biological."
7. Thompson, 385. As I show in the next chapter, Nietzsche does in fact fit into this camp.
8. Ibid., 385–86.
9. It is no accident that the second part of this course contains a detailed exploration of the concept of life and animal being, since Heidegger characterizes nihilism in part as a symptom of a crisis in the relationship between life and spirit.
10. Krell, 25.
11. Though, again, Heidegger does break from Kant in important ways, especially in ceasing to inquire after the transcendental conditions of experience.

12. Martin Heidegger, "What Is Metaphysics?," in *Basic Writings*, ed. David Farrell Krell (New York: Harper and Row, 1977,) 96 (*GA 9*).
13. Ibid., 99.
14. Ibid., 103.
15. Ibid.
16. Ibid., 109.
17. Ibid., 99.
18. *FCM*, 65.
19. *FCM*, 73. Note that one of the symptoms of spiritual sickness, nihilism, is the transition from agrarian to industrial modes of production.
20. *FCM*, 71.
21. *FCM*, 76.
22. *FCM*, 73.
23. *FCM*, 75.
24. *FCM*, 77.
25. *FCM*, 177.
26. Buchanan, 62.
27. *FCM*, 194. Here we can distinguish between a hierarchical view of the way things are, that is, things that are based upon or depend upon one another, and a hierarchical value-system, that one level or grade is better than another.
28. The entry on "Great Chain of Being" can be found at http://www.britannica.com/EBchecked/topic/243044/Great-Chain-of-Being.
29. *FCM*, 193.
30. *FCM*, 194.
31. *FCM*, 195.
32. *FCM*, 180.
33. *FCM*, 189.
34. *FCM*, 222.
35. *FCM*, 229.
36. *FCM*, 238.
37. Krell, 127.
38. McNeill, 45.
39. *FCM*, 264.
40. *FCM*, 262.
41. *FCM*, 264. Again, notice the Lamarckian strain.
42. Ted Toadvine, *Merleau-Ponty's Philosophy of Nature* (Evanston: Northwestern University Press, 2009), 52.
43. Cited by Krell, 17.
44. Martin Heidegger, *The Metaphysical Foundations of Logic* (Bloomington: Indiana University Press, 1984), 210 (*GA 26*).
45. See, for instance, Heidegger's remarks on Friedrich Gottl, *Die Grenzen der Geschichte* (Leipzig: Ducker and Humblot, 1904) in *Being and Time*.
46. The essay can be found online at: http://www.stanford.edu/dept/relstud/Sheehan/pdf/45%201983%20ON%20THE%20WAY%20TO%20EREIGNIS.pdf.

47. Daniel C. Dennett, *Darwin's Dangerous Idea: Evolution and the Meanings of Life* (New York: Simon and Schuster, 1995).
48. Moore, 27.
49. Schalow, 456.
50. Ibid., 461.
51. Ibid., 447.
52. Ibid.
53. Cited by Schalow, 457.
54. Ibid., 457. As we'll see, Nietzsche has a similar objection.
55. Buchanan, 8.
56. Ibid., 17. Buchanan correctly notes that von Uexküll's reading of Darwin is dubious.
57. Jonas, 2.
58. See Charles Bambach, *Heidegger's Roots: Nietzsche, National Socialism, and the Greeks* (Ithaca: Cornell University Press, 2003), and John D. Caputo, *Demythologizing Heidegger* (Bloomington: Indiana University Press, 1993).
59. Kelly Oliver, "Strange Kinship: Heidegger and Merleau-Ponty on Animals," *Epoche* 7, no. 1 (2008): 103.
60. Cited by Oliver, 103–04.
61. Ibid., 104.
62. Ibid.
63. Krell, 127.
64. Bruce Foltz, *Inhabiting the Earth: Heidegger, Environmental Ethics, and the Metaphysics of Nature* (New Jersey: Humanities Press, 1995), 131.
65. Ibid., 132.
66. Tristan Moyle, "Re-enchanting Nature: Human and Animal Life in Later Merleau-Ponty," *Journal of the British Society for Phenomenology* 38, no. 2 (May 2007): 164.
67. Cited by McNeill, 37.
68. Ibid.
69. Ibid., 38.
70. Foltz, 132–33.
71. In fact, Foltz's brief sketch of nature, life, and psyche more closely resembles Nietzsche's.
72. Didier Franck, "Being and the Living," in *Who Comes After the Subject?*, ed. Cadava, Connor, and Nancy (New York: Routledge, 1991), 146.
73. Ibid., 105.
74. Giorgio Agamben, *The Open: Man and Animal*, trans. Kevin Attell (Stanford: Stanford University Press, 2004), 39.
75. Cited by Krell, 113.
76. Jacques Derrida, *Of Spirit: Heidegger and the Question*, trans. Geoffrey Bennington and Rachel Bowlby (Chicago: University of Chicago Press, 1989), 55.
77. *FCM*, 13.

78. Philippe Huneman, "From the *Critique of Judgment* to the Hermeneutics of Nature: Sketching the Fate of Philosophy of Nature after Kant," *Continental Philosophy Review* 39 (2006): 1–34. For a history of the concept of nature that reaches deeper into the Western tradition, see Pierre Hadot, *The Veil of Isis: an Essay on the History of the Idea of Nature* (Cambridge: Belknap Press, 2006).

79. Huneman, 12.

80. Ibid., 6.

81. Ibid., 5.

82. Robert J Richards, *The Romantic Conception of Life: Science and Philosophy in the Age of Goethe* (Chicago: University of Chicago Press, 2002), 66.

83. Ibid., 67.

84. Cited by Richards, 242n.12, 237n.86.

85. Cited by Richards, 237n.87.

86. Richards, 290.

87. Ibid.

Chapter 5

1. Haar, 5.

2. Charles Bambach, *Heidegger's Roots: Nietzsche, National Socialism, and the Greeks* (Ithaca: Cornell University Press, 2003), introduction and chapters 4 and 5.

3. John Caputo, *Demythologizing Heidegger* (Bloomington: Indiana University Press), 1993.

4. These critiques will be discussed in detail in the next chapter in order to highlight how problems in Heidegger's accounts of history and nihilism affect his view of nature.

5. Haar, 57–63.

6. There is a curious parallel here with Kant's third critique: Kant left aesthetic judgment and a nonscientific conception of nature out of the transcendental analytic, much like Heidegger left them out of his existential analytic.

7. Haar, 6.

8. Parkes, 142 n. 26.

9. Martin Heidegger, "The Origin of the Work of Art," in *Heidegger: Basic Writings*, ed. David Farrell Krell (New York: HarperCollins, 1977), 168 (*GA 5*). Hereafter abbreviated "Origin" and "*Basic Writings*."

10. Michael Zimmerman and Sean Esbjörn-Hargens, *Integral Ecology: Uniting Multiple Perspectives on the Natural World* (New York: Integral Books, 2009), 164.

11. Ibid., 159.

12. Peter Bowler, *The Norton History of the Environmental Sciences* (New York: W.W. Norton and Company, 1992), 522.

13. *Integral Ecology*, 161.

14. Quoted by Bowler, 530.

248 / Notes to Chapter 5

15. Ibid., 162.
16. Bowler, 538.
17. *Integral Ecology*, 168.
18. Ibid., 165.
19. Bowler, 540.
20. This is driven home, for instance, in the 1951 essay "The Thing." See Martin Heidegger, "The Thing," in *Poetry, Language, Thought*," trans. Albert Hofstadter (New York: HarperCollins, 1971) (*GA* 7). The proper attitude toward things, "releasement" (Gelassenheit), is articulated in Martin Heidegger, *Discourse on Thinking* (New York: Harper and Row, 1966) (*GA* 77). I will return to these themes later in this chapter.
21. Heidegger writes that "essential reflection upon technology and decisive confrontation with it must happen in a realm that is, on the one hand, akin to the essence of technology and, on the other, fundamentally different from it. Such a realm is art." Martin Heidegger, "The Question concerning Technology," in *The Question concerning Technology and Other Essays*, trans. and ed. William Lovitt (New York: Harper and Row, 1977), 35 (*GA* 7).
22. "Origin," 210.
23. Ibid., 148–49.
24. Ibid., 151.
25. Ibid., 152.
26. Ibid.
27. Ibid.
28. Ibid., 154.
29. Julian Young, *Heidegger's Later Philosophy* (New York: Cambridge University Press, 2002), 10.
30. Ibid., 9.
31. "Origin," 160.
32. *Being and Time*, 100.
33. Ibid., 172.
34. *Being and Time*, 231.
35. Jacques Taminiaux, *Heidegger and the Project of Fundamental Ontology* (Albany: SUNY Press, 1991), 128.
36. "Origin," 170.
37. Ibid., 167.
38. Ibid., 168.
39. Ibid., 173.
40. Ibid.
41. Ibid., 188.
42. Martin Heidegger, *Aristotle's* Metaphysics Theta 1–3 *On the Essence and Actuality of Force*, trans. Walter Brogan and Peter Warnek (Bloomington, Indiana University Press, 1995) (*GA* 33).
43. Haar, 108.

44. "Origin," 195.

45. It is strange, though, that Heidegger did not parlay this insight that "art completes nature" into a more biological notion that human beings are the continuation and extension of a process of life developing in and as nature.

46. Ibid., 60.

47. Ibid., 170.

48. As I noted in the previous chapter, in *Fundamental Concepts of Metaphysics*, he does entertain a scalar model of humans, animals, and inorganic beings in terms of worldliness.

49. "Letter on Humanism," 228.

50. Caputo, 126.

51. Martin Heidegger, *Introduction to Metaphysics,* trans. Ralph Mannheim (New York: Yale University Press, 1959), 11–12 (*GA* 40).

52. Quoted by Haar, 11.

53. *Introduction to Metaphysics*, 59.

54. Ibid., 85.

55. Ibid., 12–13.

56. Ibid., 14.

57. But the capital point has to do with Heidegger's development: the fundamental ontology of *Being and Time*, in which he said that we must above all refrain from telling a story—in other words, refrain from metaphysics—has been supplanted by a history of Being, a most magisterial master narrative, an audacious account, a tall tale—one is tempted to say, a conspiracy theory—about the Western tradition. It is a story of decline, of "world-darkening."

58. Ibid., 51.

59. As Haar points out, Heidegger's reconsideration of *physis* is spurred in no small part by the writings of Heraclitus. See Haar, 49–52.

60. Again, however, we must heed the political context: this is not long after Heidegger's jeremiad on the self-assertion of the German University of 1933, in which the more individualistic notions of anticipatory resoluteness, authenticity, and choosing one's hero from *Being and Time* are focused toward and fused with the struggle to align the many wills with the one great will of the führer to achieve the national destiny.

61. See Bambach, 162, and Michael E. Zimmerman, "The Ontological Decline of the West," in *A Companion to Heidegger's* Introduction to Metaphysics, ed. Richard Polt and Gregory Fried (New Haven: Yale University Press, 2001).

62. Martin Heidegger, "On the Essence and Concept of *Physis* in Aristotle's Physics B," trans. Thomas Sheehan in *Pathmarks*, ed. William McNeill (New York: Cambridge University Press, 1998), 185 (*GA* 9).

63. See the lecture course cited above as well as Martin Heidegger, *Elucidations of Hoelderlin's Poetry*, trans. Keith Hoeller (New York: Humanity Books, 2000) (*GA* 4).

64. Ibid., 183.

65. For a detailed discussion of Jünger's influence on Heidegger, see Michael E. Zimmerman, *Heidegger's Confrontation with Modernity* (Bloomington: Indiana University Press, 1990), chapters 4–6.
66. *Pathmarks*, 184.
67. Ibid., 191.
68. Ibid., 208.
69. Ibid.
70. Ibid., 210.
71. Ibid., 214.
72. Ibid., 136.
73. Ibid., 193. Cf. Martin Heidegger, "The Question concerning Technology," in *The Question concerning Technology and Other Essays*, trans. and ed. William Lovitt (New York: Harper and Row, 1977), 8 (*GA 7*). Hereafter abbreviated "QCT."
74. Ibid., 217.
75. Ibid.
76. Quoted by Neil Evernden, *The Natural Alien* (Buffalo: University of Toronto Press, 1985), 53.
77. *Pathmarks*, 226.
78. Ibid., 226–27.
79. Ibid., 195.
80. A full account of Heidegger's view of technology needs to show how it fits within his narrative of the history of being. This will be taken up in the next chapter.
81. Martin Heidegger, *Contributions to Philosophy (From Enowning)*, trans. Parvis Emad and Kenneth Maly (Bloomington: Indiana University Press, 1999), 92 (*GA 65*).
82. Ibid., 88.
83. Ibid., 92.
84. Martin Heidegger, *Mindfulness*, trans. Parvis Emad and Thomas Kalary (New York: Continuum, 2006), 19 (*GA 66*).
85. *Contributions*, 86.
86. Ibid., 84.
87. Ibid., 86.
88. *Mindfulness*, 170.
89. "QCT," 5.
90. "QCT," 11.
91. "QCT," 14.
92. "QCT," 20.
93. "QCT," 27.
94. C. S. Lewis, *The Abolition of Man* (New York: HarperCollins, 1944), 68.
95. "QCT." 27.
96. "QCT."
97. Young, 77.
98. QCT," 34.
99. Quoted by Young, 64.

100. Again, while in *Being and Time* all of this is discussed with reference to Dasein, the conclusion is that Dasein is care, which implies something to be cared *for*; though the latter is not fleshed out in the text, this is not grounds for claiming that Heidegger underwent some momentous transformation in his later work.

101. Martin Heidegger, "The Thing," in *Poetry, Language, Thought,* trans. Albert Hofstadter (New York: HarperCollins, 1971), 167 (*GA 7*).

102. Ibid., 167.

103. Quoted by Young, 99.

104. Ibid., 94.

105. Charles Taylor, *A Secular Age* (Cambridge: Belknap Press, 2008).

106. As Young points out, and as Taminiaux and others have documented, Heidegger's inspiration here is Nietzsche: "[T]he idea of the exemplary 'hero,' the 'role-' or, better, 'life-model,' is taken from Nietzsche's 'On the Uses and Abuses of History,' from specifically, his conception of the 'memorializing' function of history." Young, 97n.8.

107. Bret Davis, *Heidegger and the Will: On the Way to Gelassenheit* (Evanston: Northwestern University Press, 2007).

108. Martin Heidegger, *Discourse on Thinking* trans. John Anderson and E. Hans Freund (New York: Harper and Row, 1966), 53 (*GA 77*).

109. Ibid., 46.

110. *Contesting Earth's Future*, 118. While Jonas perhaps goes too far in suggesting Heidegger espoused a hostile cosmos, he is correct that the very notion of a cosmos is never fully addressed in Heidegger's thought.

111. John Van Buren makes a similar argument in the final chapter of *The Young Heidegger*. For a summary of the Christian stewardship ethic, see Robin Attfield, "Stewardship versus Exploitation," in Donald VanDeVeer and Christine Pierce, *The Environmental Ethics and Policy Book* (Belmont: Wadsworth/Thomson, 2003), 66–70. For a defense, see John Passmore, *Man's Responsibility for Nature* (London: Duckworth, 1974).

Chapter 6

1. It is noteworthy that his own essay on art, which advanced the ontological primacy of *poiesis* over *techne*, was composed around just this time.

2. Haar, 79.

3. Ibid., 2.

4. See Charles Bambach's warning about a naïve ecological reading of Heidegger's earth below.

5. Haar, 3.

6. Ibid., 68.

7. Ibid., 72.

8. Ibid., 82.

9. Bret W. Davis, *Heidegger and the Will: On the Way to Gelassenheit* (Evanston: Northwestern University Press, 2007), 15.
10. Hereafter abbreviated *IM*.
11. Davis, chapter 3.
12. *IM*, 37.
13. *IM*, 31–32.
14. Stanley Rosen, *The Question of Being: A Reversal of Heidegger* (New Haven: Yale University Press, 1993), xviii–xix.
15. *IM*, 49–50.
16. John D. Caputo, *Demythologizing Heidegger* (Bloomington: Indiana University Press, 1993), 119.
17. Ibid.
18. Caputo points out how this viral logic affects Heidegger's understanding of human and animal life and how his attempt to avoid traditional essentialism backfires: "Heidegger reproduces all the essentials of essentialism by clinging to the distinction between the pure inside of human being—where there is truth, clearing, Being, language world—and the impure, contaminated outside—where there are only brute stupidity, mute silence, in a word, mere life. Inside and outside are separated by an abyss" (125). His treatment of animals, Caputo writes, "verges on a Cartesianism which treats them as little more than machines" (125). In *IM*, then, the distinction—or dissociation—between "spirit" and "animal" or "spirit" and "life" is at its greatest.
19. See chapter 4.
20. Michael E. Zimmerman, "The Ontological Decline of the West," online resource: http://www.colorado.edu/ArtsSciences/CHA/profiles/zimmpdf/introd_to_meta.pdf, 10.
21. Ibid., 10–11.
22. Ibid., 14.
23. Ibid., 17.
24. Ibid., 18.
25. Caputo, 2.
26. Ibid., 10.
27. Ibid., 17.
28. Ibid., 4.
29. Quoted by Caputo, 210.
30. Ibid., 118.
31. Ibid., 285.
32. Ibid., 5.
33. Alan Schrift, "Nietzssche's Psycho-Genealogy: A Ludic Alternative to Heidegger's Reading of Nietzsche," *Journal of the British Society for Phenomenology* 14, no. 3 (October 1983): 283.
34. Laurence Lampert, "Heidegger's Nietzsche Interpretation," *Man and World: An International Philosophical Review* 7 (November 1974): 353.

35. Walter Kaufmann, *Discovering the Mind, Volume Two: Nietzsche, Heidegger, and Buber* (New York: McGraw-Hill, 1980), 172.

36. David Farrell Krell, "Heidegger's Reading of Nietzsche: Confrontation and Encounter," *Journal of the British Society for Phenomenology* 14, no. 3 (October 1983): 273.

37. Jacques Taminiaux, "On Heidegger's Interpretation of the Will to Power as Art," *New Nietzsche Studies*, vol. 3, nos. 1–2 (Winter 1999): 7.

38. Ibid., 3.

39. Ibid., 15.

40. D. W. Conway, "Heidegger, Nietzsche, and the Origins of Nihilism," *Journal of Nietzsche Studies* 3 (1992): 11–43.

41. Lampert, 356.

42. Friedrich Nietzsche, *Beyond Good and Evil*, trans. Walter Kaufmann (New York: Vintage Books, 1966), 31–32.

43. Lampert, 361.

44. Ibid.

45. Ibid.

46. Martin Heidegger, *Nietzsche: Volume 4, Nihilism*, trans. David Farrell Krell (New York: HarperCollins, 1982), 27–28 (*GA 6*). Hereafter abbreviated *Nietzsche*.

47. Ruth Irwin, "Heidegger and Nietzsche: The Question of Value and Nihilism in Relation to Education, *Studies in Philosophy and Education* 22, no. 3–4 (May–July 2003): 283.

48. *Beyond Good and Evil*, 47–48.

49. Ibid., 48.

50. Ibid.

51. See Parkes' introduction to Friedrich Nietzsche, *Thus Spoke Zarathustra*, trans. Graham Parkes (New York: Oxford University Press, 2004), xx. Schopenhauer's position on the renunciation of suffering and the primordial unity of all beings has been enlisted in the defense of animal rights. See Gary E. Varner, "The Schopenhauerian Challenge in Environmental Ethics," *Environmental Ethics* 7 (1985): 209–30.

52. Ibid., xx.

53. Walter Kaufmann, *Nietzsche: Philosopher, Psychologist, Antichrist*, 4th ed. (Princeton: Princeton University Press, 1974), 239.

54. Ibid., 238.

55. Lawrence Vogel, "Hans Jonas's Diagnosis of Nihilism: The Case of Heidegger," *International Journal of Philosophical Studies* 3, no. 1 (March 1995): 58.

56. Davis, 152.

57. Ibid., 24.

58. Ibid., 307.

59. Ibid., xxx.

60. *Beyond Good and Evil*, 21.

61. Davis, 150.

62. Quoted by Davis, 175.
63. *Nietzsche*, 27.
64. Lampert, 364.
65. Julian Young, "Being and Value: Heidegger Contra Nietzsche," *International Studies in Philosophy* 27, no. 3 (1995): 100.
66. See the section "On the 1,001 Goals," in Friedrich Nietzsche, *Thus Spoke Zarathustra*, trans. Graham Parkes (New York: Oxford University Press, 2004).
67. Lampert, 365.
68. Ibid., 365, my italics.
69. Krell, 276.
70. *Nietzsche*, 62.
71. Parvis Emad, "Heidegger's Value-Criticism and Its Bearing on the Phenomenology of Values," *Research in Phenomenology* 7 (1977): 190.
72. Martin Heidegger, Nietzsche, *Volume One: The Will to Power as Art*, trans. David Farrell Krell (New York: HarperCollins, 1982), 10 (*GA* 43).
73. Ibid., 50.
74. Dick White, "Heidegger on Nietzsche on the Question of Value," in *Postmodernism and Continental Philosophy*, ed. Hugh Silverman (Albany: SUNY Press, 1988), 114.
75. "Heidegger's Nietzsche-Interpretation," 4.
76. Martin Heidegger, *What Is Called Thinking?*, trans J. Glenn Gray (New York: Harper and Row, 1976), 69 (*GA* 8).
77. White, 114.
78. Ibid., 112.
79. Ibid., 115.
80. Ibid., 118.
81. Krell, 276.
82. *Nietzsche*, 129.
83. Ibid.
84. Quoted by Heidegger, 135.
85. Quoted by Parkes, 1.
86. Ibid.
87. Schrift, 289.
88. *Beyond Good and Evil*, 23.
89. Ibid., 24.
90. "Heidegger's Nietzsche-Interpretation," 9.
91. Michael E. Zimmerman, *Heidegger's Confrontation with Modernity* (Bloomington: Indiana University Press, 1990), 124, my italics.

Chapter 7

1. R. J. Hollingdale, *Nietzsche: The Man and His Philosophy* (London: Routledge and Kegan Paul, 1965), 90.

2. Tamler Sommers and Alex Rosenberg, "Darwin's Nihilistic Idea: Evolution and the Meaninglessness of Life," *Biology and Philosophy* 18 (2003): 653.

3. Christopher Janaway and Simon Robertson, eds., *Nietzsche, Naturalism, and Normativity* (Oxford: Oxford University Press 2012).

4. Graham Parkes, *Composing the Soul: Reaches of Nietzsche's Psychology* (Chicago: University of Chicago Press, 1994).

5. *Nietzsche, Naturalism, and Normativity*, 238.

6. Ibid., 239.

7. Ibid., 238.

8. Ibid., 241.

9. Michael E. Zimmerman, "Nietzsche and Ecology: A Critical Inquiry," available online at http://www.colorado.edu/ArtsSciences/CHA/profiles/zimmpdf/nietz_and_ecology.pdf, 16.

10. Ted Benton, "Naturalism in Social Science," in *Routledge Encyclopedia of Philosophy*, ed. E. Craig (London: Routledge, 1998), retrieved August 8, 2010, from http://www.rep.routledge.com/article/RO11.

11. See, e.g., Friedrich Nietzsche, *The Will to Power*, trans. Walter Kaufmann (New York: Vintage Books, 1967), sections 647 and 685. Hereafter abbreviated *WP*.

12. Gregory Moore, *Nietzsche, Biology, Metaphor* (New York: Cambridge University Press, 2002), 71.

13. Friedrich Nietzsche, *The Gay Science*, trans. Walter Kaufmann (New York: Vintage Books, 1974), 335. Hereafter *GS*.

14. Ibid., 335.

15. Moore, 3.

16. Ibid., 26.

17. William Bechtel and Robert Richardson, "Vitalism," in *Routledge Encyclopedia of Philosophy* (London: Routledge, 1998), available online at http://mechanism.ucsd.edu/teaching/philbio/vitalism.htm.

18. Ibid., 53.

19. Daniel Dennett, *Darwin's Dangerous Idea: Evolution and the Meanings of Life* (New York: Simon and Schuster, 1995), 466.

20. John Richardson, *Nietzsche's New Darwinism* (New York: Oxford University Press, 2004), 5.

21. Ibid., 12.

22. Ibid., 6–7.

23. Ibid., 15.

24. Nietzsche's views on Buddhism are another example of this pattern. While he attacks Buddhism for its purported pessimism and life-denying spirit, his knowledge of it was at best limited and at worst wrong, and his own mature views have a deep affinity with much Buddhist thought, especially the Mahayana tradition. See Graham Parkes, ed., *Nietzsche and Asian Thought* (Chicago: University of Chicago Press, 1991), introduction.

25. Moore, 63.

26. Richardson, 17.

27. *WP*, 332.
28. *WP*, 365.
29. Moore, 37–39.
30. Ibid., 83.
31. *WP*, 341.
32. *WP*, 340.
33. *WP*, 342.
34. *WP*, 20.
35. Friedrich Nietzsche, *Beyond Good and Evil*, trans. Walter Kaufmann (New York: Vintage Books, 1964), 21, my emphasis. Hereafter abbreviated *BGE*.
36. Cited by Richardson, 20.
37. Richardson, 21.
38. Ibid., 22–23.
39. *WP*, *378*.
40. *WP*, 339, my emphasis.
41. *WP*, 344.
42. Moore, 29.
43. Robert Richards, *The Romantic Conception of Life: Science and Philosophy in the Age of Goethe* (Chicago: University of Chicago Press, 2002), 514.
44. Ibid., 534.
45. Dennett, 126.
46. Richards, 66.
47. Ibid., 237.
48. Ibid., 67.
49. Cited by Richards, 242n.12, 237n.86.
50. Cited by Richards, 237n.87.
51. Evan Thompson, *Mind in Life: Biology, Phenomenology, and the Sciences of Mind* (Cambridge: Harvard University Press, 2007), 140.
52. Richards, 130.
53. Ibid.
54. Ibid., 295.
55. *WP*, 342.
56. *WP*, 346.
57. Cited by Richards, 306.
58. Cited by Richards, 299.
59. Richards, 297.
60. *WP*, 338–39.
61. Thompson, 70.
62. *WP*, 381.
63. *WP*, 347.
64. Jonas, 16.
65. Ibid.
66. Ibid., 20.

67. Ibid., 57.
68. Ibid., 86.
69. Thompson, 44.
70. Ibid., 147.
71. Jonas, 79.
72. Thompson, 164.
73. Jonas, 58.
74. Ibid., 2.
75. Ibid., 282.
76. Ibid., 139. Note that this view of the organism is quite similar to that offered by Heidegger in the 1929–30 lecture course. However, as we saw in previous chapters, Heidegger fails to integrate this with his view of humans and his philosophy of nature.
77. Jonas, 140.
78. Ibid., 147.
79. Cited by Thompson, 154.
80. Richardson, 7.
81. Ibid., 26.
82. *WP*, 344, my emphasis.
83. *WP*, 375–76.
84. Cited by Moore, 39.
85. Ibid., 40.
86. Ibid.
87. Graham Parkes, *Composing the Soul: Reaches of Nietzsche's Psychology* (Chicago: University of Chicago Press, 1994), 359.
88. Richardson, 34.
89. Ibid., 42.
90. Cited by Richardson, 33n.64.
91. Ibid., 55.
92. Ibid., 37.
93. Ibid., 63–64.
94. Cited by Krell, 17.
95. Cobb, ix.
96. In Cobb, 223.
97. Ibid., 228.
98. Ibid., 216.
99. Ibid., 218.
100. Cited by Richardson, 72.
101. Ibid.
102. Ibid., 73.
103. Ibid.
104. Ibid., 72.
105. Ibid.

106. Ibid., 74.

107. This view also closely resembles Heidegger's pre-*Being and Time* ontology of life. See Buchanan, chapter 4 for Uexkull's influence on Merleau-Ponty.

108. Thompson, 71.

109. Cited by Richardson, 74. In the following chapter, I discuss the similarity of this view to Paul Taylor's biocentrism.

110. Richardson, 76.

111. Richards, 516.

112. Ibid., 553.

113. Moore, 30.

114. Richardson, 71.

115. Ibid., 85.

116. Ibid., 83.

117. Richardson alleges that Nietzsche's notion of social selection prefigures the "meme" theory popularized by Dawkins, according to which certain moral, cultural, and religious worldviews are understood as self-replicating entities, like genes, that are transmitted through individuals but not by individuals.

118. Richardson, 84.

119. D. W. Conway, "Heidegger, Nietzsche, and the Origins of Nihilism," *Journal of Nietzsche Studies* 3 (1992): 11–43.

120. Richardson, 113.

121. Ibid., 125.

122. Ibid., 130.

123. Ibid.

Chapter 8

1. See *New Nietzsche Studies* 5, nos. 1–2 (Spring/Summer 2002), which is devoted to Nietzsche and ecology.

2. "Nietzsche and Ecology," 2.

3. Max Hallman, "Nietzsche's Environmental Ethics," *Environmental Ethics* 13 (Summer 1991): 100–01.

4. Quoted by Hallman, 110.

5. Hallman, 116.

6. Ibid., 115.

7. See my discussion of biocentrism and deep ecology in chapter 1.

8. Quoted by Hallman, 115.

9. Hallman, 117.

10. Ibid., 123.

11. See Michael E. Zimmerman and Sean Esbjorn-Hargens, *Integral Ecology: Uniting Multiple Perspectives on the Natural World*, Boston: Integral Books, 2009), 6, 479. They define "subtle reductionism" as "the reduction of all interiors to interob-

jective phenomena (reducing the 'I' and 'we' perspectives to interwoven systems—'its'). The science of ecology has typically exemplified [this]," 6.

12. At *WP* 711, Nietzsche says that "the world is not an organism at all," 379.
13. *GS*, 167.
14. Ralph Acampora, "Using and Abusing Nietzsche for Environmental Ethics," *Environmental Ethics* 16 (Summer 1994): 189.
15. Ibid., 193.
16. Martin Drenthen, "Nietzsche and the Paradox of Environmental Ethics: Nietzsche's View of Nature and Morality," *New Nietzsche Studies* 5, nos. 1–2 (2002): 14.
17. Ibid., 15.
18. Ibid.
19. Ibid., 20.
20. Ibid., 21.
21. Ibid., 22.
22. Ibid., 18. Drenthen is referring to Bataille, Foucault, Deleuze, and Lyotard. For an overview of Nietzsche's influence on the French, see Alan Schrift, *Nietzsche's French Legacy* (New York: Routledge, 1995). Also see Michael Soule, *Reinventing Nature? Responses to Postmodern Deconstruction* (Washington, D.C.: Island Press, 1995).
23. See Neil Evernden, *The Social Creation of Nature* (Baltimore: Johns Hopkins University Press, 1992).
24. Ted Toadvine, *Merleau-Ponty's Philosophy of Nature* (Evanston: Northwestern University Press, 2009), 13.
25. Steven Vogel, *Against Nature: The Concept of Nature in Critical Theory* (Albany: SUNY Press, 1996), 6, my italics.
26. For a recent example of alterity theory, see the conclusion to Toadvine's book cited above.
27. *BGE*, 47.
28. *BGE*, 48.
29. *BGE*, 48.
30. Graham Parkes, "Staying Loyal to the Earth," in *Nietzsche's Futures*, ed. John Lippitt (London: Macmillan Press, 1999), 168.
31. Ibid.
32. Quoted by Parkes, 170.
33. *GS* 169. This is where Nietzsche speaks of "naturalizing" humanity.
34. *BGE*, 161.
35. Ibid., 89.
36. Parkes, 170.
37. Ibid., 179.
38. Graham Parkes, "Nietzsche's Environmental Philosophy: A Trans-European Perspective," *Environmental Ethics* 27 (Spring 2005): 84.
39. Ibid.

40. Ibid., 85.
41. Ibid., 89.
42. Ibid., 176.
43. "Nietzsche's Environmental Philosophy," 85.
44. *BGE*, 148, my emphasis.
45. See, for instance, Tom Regan, *The Case for Animal Rights* (Berkeley: University of California Press, 1983), 361–62, and Luc Ferry, *The New Ecological Order*, trans. Carol Volk (Chicago: University of Chicago Press, 1995), for critiques of the eco-fascist element in deep ecology.
46. Paul W. Taylor, *Respect for Nature: A Theory of Environmental Ethics* (Princeton: Princeton University Press, 1986).
47. Walter Kaufmann, *Nietzsche: Philosopher, Psychologist, Antichrist* (Princeton: Princeton University Press, 1974), 280.
48. Ibid., 231.
49. Erazim Kohak, *The Embers and the Stars* (Chicago: University of Chicago Press, 1984), 55.
50. Ibid., 98.
51. Ibid., 99.
52. *Integral Ecology*, 104.
53. Ibid., 11.
54. *Nietzsche and Ecology*, 18.
55. David Ray Griffin, "Whitehead's Deeply Ecological Worldview," in *Worldviews and Ecology*, ed. Mary Evelyn Tucker and John A. Grim (London: Bucknell University Press, 1993), 192–93.
56. Ibid., 203.
57. For instance, see Donald VanDeVeer, "Interspecific Justice," *Inquiry: An Interdisciplinary Journal of Philosophy* 24, nos. 1–4 (1979): 55–79.

Bibliography

Abram, David. *The Spell of the Sensuous*. New York: Pantheon Books, 1996.
Acampora, Ralph. "Using and Abusing Nietzsche for Environmental Ethics." *Environmental Ethics* 16 (Summer 1994): 87–194.
Agamben, Giorgio. *The Open: Man and Animal*, trans. Kevin Attell. Stanford: Stanford University Press, 2004.
Armstrong, Susan, and Richard Botzler, eds. *Environmental Ethics: Divergence and Convergence*. Boston: McGraw-Hill, 1993.
Agar, Nicholas. *Life's Intrinsic Value: Science, Ethics, and Nature*. New York: Columbia University Press, 2001.
Avery, Stephen. "The Misbegotten Child of Deep Ecology," *Environmental Values* 13, no. 1 (February 2004): 31–50.
Babich, Babette. "Heidegger's Relation to Nietzsche's Thinking: Connivance, Nihilism, and Value." *New Nietzsche Studies* 3, nos. 1–2 (Winter 1999).
Badiner, Allan Hunt, ed. *Dharma Gaia: A Harvest of Essays in Buddhism and Ecology*. Berkeley: Parallax Press, 1990.
Bambach, Charles. *Heidegger's Roots: Nietzsche, National Socialism, and the Greeks*. Ithaca: Cornell University Press, 2003.
Barnhill, David Landis. *Deep Ecology and World Religions: New Essays on Sacred Grounds*. Albany: SUNY Press, 2001.
Bechtel, William, and Robert Richardson. "Vitalism," In *Routledge Encyclopedia of Philosophy*. London: Routledge, 1998.
Benton, Ted. "Naturalism in Social Science." In *Routledge Encyclopedia of Philosophy*, ed. E. Craig (London: Routledge, 1998): retrieved August 8, 2010, from http://www.rep.routledge.com/article/RO11.
Botha, Catherine Frances. "Heidegger, Technology, and Ecology." *South African Journal of Philosophy* 22, no. 2 (2003): 157–72.
Bowler, Peter. *The Norton History of the Environmental Sciences*. New York: W. W. Norton and Company, 1992.
Brown, Charles S., and Ted Toadvine, eds. *Eco-Phenomenology: Back to the Earth Itself*. Albany: SUNY Press, 2003.
Buchanan, Brett. *Onto-Ethologies: The Animal Environments of Uexkull, Heidegger, Merleau-Ponty, and Deleuze*. Albany: SUNY Press, 2008.

Burtt, E. A. *The Metaphysical Foudations of Modern Science*. New York: Doubleday Anchor Books, 1954.
Callicott, J. Baird. "Environmental Ethics." In *Encyclopedia of Applied Ethics*, ed. Dan Callahan (Academic Press, 1998): 311.
Cameron, W. S. K. "Heidegger's Concept of the Environment in *Being and Time*." *Environmental Philosophy* 1, no. 1 (Spring 2004): 34–46,.
Campbell, Keith. "Naturalism," In *Encyclopedia of Philosophy*, ed. Donald Borchert 6, 492–95.
Campora, Ralph A. "Using and Abusing Nietzsche for Environmental Ethics" *Environmental Ethics* 16 (1994): 87–194.
Caputo, John D. *Against Ethics*. Bloomington: Indiana University Press, 1993.
———. *Demythologizing Heidegger*. Bloomington: Indiana University Press, 1993.
Cobb, Jr., John, ed. *Back to Darwin: A Richer Account of Evolution*. Grand Rapids: William B. Eerdmans Publishing Company, 2008.
Colony, Tracy. "Before the Abyss: Agamben on Heidegger and the Living." *Continental Philosophy Review* 40 (2007).
———. "Dwelling in the Biosphere? *Heidegger's* Critique of Humanism and Its Relevance for Ecological Thought." *International Studies in Philosophy* 31, no. 1 (1999): 37–45.
Conway, D. W. "Heidegger, Nietzsche, and the Origins of Nihilism." *Journal of Nietzsche Studies* 3 (1992): 11–43.
Cooper, David. "Heidegger on Nature." *Environmental Values* 14, no. 3 (August 2005): 339–51.
Craig, William Lane, and J. P. Moreland. *Naturalism: A Critical Analysis*. New York: Routledge, 2000.
Crosby, Donald A. *The Specter of the Absurd: Sources and Criticisms of Modern Nihilism*. Albany: SUNY Press, 1988.
Davis, Bret W. *Heidegger and the Will: On the Way to Gelassenheit*. Evanston: Northwestern University Press, 2007.
De Caro, Mario, and David Macarthur, eds. *Naturalism in Question*. Cambridge: Harvard University Press, 2004.
DeLuca, Kevin Michael. "Thinking with Heidegger: Rethinking Environmental Theory and Practice." *Ethics and the Environment* 10, no. 1 (Spring 2005): 67–87.
Dennett, Daniel C. *Darwin's Dangerous Idea: Evolution and the Meanings of Life*. New York: Simon and Schuster, 1995.
Derrida, Jacques. *Of Spirit: Heidegger and the Question*, trans. Geoffrey Bennington and Rachel Bowlby. Chicago: University of Chicago Press, 1989.
Devall, Bill, and George Sessions. *Deep Ecology: Living as if Nature Mattered*. Salt Lake City: Peregrine Smith Books, 1985.
Dombrowski, Daniel. "Heidegger's Anti-Anthropocentrism." *Between the Species* 10, nos. 1–2 (1994): 26–38.
Drenthen, Martin. "Nietzsche and the Paradox of Environmental Ethics." *New Nietzsche Studies* 5, nos. 1–2 (Spring–Summer 2002): 12–25.

Dreyfus, Hubert. *Being-in-the-World: A Commentary of Heidegger's* Being and Time, *Division I.* Cambridge: MIT Press, 1991.

Elden, Stuart. "Heidegger's Animals." *Continental Philosophy Review* 39 (2006): 273–91.

Emad, Parvis. "Heidegger's Value-Criticism and Its Bearing on the Phenomenology of Values." *Research in Phenomenology* 7 (1977): 190–208.

Evernden, Neil. *The Natural Alien.* Toronto: University of Toronto, 1993.

Ferry, Luc. *The New Ecological Order*, trans. Carol Volk. Chicago: University of Chicago Press, 1995.

Flanagan, Owen. *The Really Hard Problem: Meaning in a Material World.* Cambridge: MIT Press, 2007.

Foltz, Bruce V. *Inhabiting the Earth: Heidegger, Environmental Ethics, and the Metaphysics of Nature.* New Jersey: Humanities Press, 1995.

Foltz, Bruce, and Robert Frodeman, eds. *Rethinking Nature: Essays in Environmental Philosophy.* Bloomington: Indiana University Press, 2004.

Franck, Didier. "Being and the Living." In *Who Comes after the Subject?*, ed. Cadava, Connor, and Nancy. New York: Routledge, 1991.

Frodeman, Robert. "The Policy Turn in Environmental Ethics," http://www.phil.unt.edu/chile/docs/frodemanEn_Ethics_05.pdf.

Gillespie, Michael. *Nihilism before Nietzsche.* Chicago: University of Chicago Press, 1995.

Glazebrook, Trish. *Heidegger's Philosophy of Science.* New York: Fordham University Press, 2001.

Grene, Marjorie. *Philosophical Approaches to Biology.* New York: Basic Books, 1968.

Griffin, David Ray. "Whitehead's Deeply Ecological Worldview." In *Worldviews and Ecology*, ed. M. E. Tucker and J. Grim. Maryknoll, NY: Orbis Books, 1994.

Guignon, Charles. *The Cambridge Companion to Heidegger.* New York: Cambridge University Press, 1993.

Haar, Michel. *The Song of the Earth: Heidegger and the Grounds of the History of Being.* Bloomington: Indiana University Press, 1993.

Hadot, Pierre. *The Veil of Isis: an Essay on the History of the Idea of Nature.* Belknap Press: Cambridge, 2006.

Hallman, Max. "Nietzsche's Environmental Ethics." *Environmental Ethics* 13, no. 2 (Summer 1991): 99–125.

Hart, James G., and Lester Embree, eds. *Phenomenology of Values and Valueing.* Boston: Kluwer Academic Publishers, 1997.

Heidegger, Martin. *Aristotle's* Metaphysics Theta 1–3 *On the Essence and Actuality of Force*, trans. Walter Brogan and Peter Warnek. Bloomington, Indiana University Press, 1995.

———. *Basic Concepts of Ancient Philosophy*, trans. Richard Rojcewicz. Bloomington: Indiana University Press, 2008.

———. *Basic Concepts of Aristotelian Philosophy*, trans. Robert D. Metcalf and Mark B. Tanzer. Bloomington: Indiana University Press, 2009.

———. *Basic Writings*, ed. David Farrell Krell. New York: Harper and Row, 1977.

———. *Being and Time*, trans. John Macquarrie and Edward Robinson. New York: Harper and Row, 1962.

———. *Contributions to Philosophy (From Enowning)*, trans. Parvis Emad and Kenneth Maly. Bloomington: Indiana University Press, 1999.

———. *Discourse on Thinking*, trans. John A. Anderson and E. Hans Freund. New York: Harper and Row, 1966.

———. *Introduction to Metaphysics*, trans. Ralph Mannheim. New York: Yale University Press, 1959.

———. *Logic: The Question of Truth*, trans. Thomas Sheehan. Bloomington: Indiana University Press, 2010.

———. *Mindfulness*, trans. Parvis Emad and Thomas Kalary. New York: Continuum, 2006.

———. *The Metaphysical Foundations of Logic*. Bloomington: Indiana University Press, 1984.

———. *Nietzsche: Volumes One and Two*, trans. David Farrell Krell. New York: Harper and Row, 1991.

———. *Nietzsche: Volumes Three and Four*, trans. David Farrell Krell. New York: Harper and Row, 1991.

———. "On the Essence and Concept of Physis in Aristotle's Physics B," trans. Sheehan. Pathmarks.

———. *On the Way to Language*, trans. Peter D. Hertz. New York: Harper and Row, 1971.

———. *Ontology—The Hermeneutics of Facticity*, trans. John van Buren. Bloomington: Indiana University Press, 1999.

———. *Pathmarks*, ed. William McNeill. Cambridge: Cambridge University Press, 1998.

———. *Phenomenological Interpretations of Aristotle*, trans. Richard Rojcewicz. Bloomington: Indiana University Press, 2001.

———. *Poetry, Language, Thought*, trans. Albert Hofstadter. New York: Harper and Row, 1971.

———. *Supplements: From the Earliest Essays to* Being and Time *and Beyond*, ed. John van Buren. Albany: SUNY Press, 2002.

———. *The End of Philosophy*, trans. Joan Stambaugh. Chicago: University of Chicago Press, 1973.

———. *The Fundamental Concepts of Metaphysics*, trans. William McNeill. Bloomington: Indiana University Press, 2001.

———. *The Question concerning Technology and Other Essays*, trans. William Lovitt. New York: Harper and Row, 1977.

———. *What Is Called Thinking?*, trans J. Glenn Gray. New York: Harper and Row, 1976.

Holland, Nancy. "Rethinking Ecology in the Western Philosophical Tradition: Heidegger and/on Aristotle." *Continental Philosophy Review* 32 (1999): 409–20.

Hollingdale, R. J. *Nietzsche: The Man and His Philosophy*. London: Routledge and Kegan Paul, 1965.
Huneman, Philippe. "From the *Critique of Judgment* to the Hermeneutics of Nature: Sketching the Fate of Philosophy of Nature after Kant." *Continental Philosophy Review* 39 (2006): 1–34.
Irwin, Ruth. "Heidegger and Nietzsche: The Question of Value and Nihilism in Relation to Education." *Studies in Philosophy and Education* 22, nos. 3–4, (May–July 2003): 227–44.
Janaway, Christopher, and Simon Robertson, eds. *Nietzsche, Naturalism, and Normativity*. Oxford: Oxford University Press, 2012.
Jonas, Hans. *The Phenomenon of Life*. New York: Harper and Row, 1966.
Kaufmann, Walter. *Discovering the Mind, Volume Two: Nietzsche, Heidegger, and Buber*. New York: McGraw-Hill, 1980.
———, ed. *Existentialism from Dostoevsky to Sartre*. New York: Penguin, 1956.
———. *Nietzsche: Philosopher, Psychologist, Antichrist*. 4th Ed. Princeton: Princeton University Press, 1974.
Keiji, Nishitani. *The Self-Overcoming of Nihilism*, trans. Graham Parkes. Albany: SUNY Press, 1990.
Kisiel, Theodore, and John van Buren, eds. *Reading Heidegger from the Start: Essays in His Earliest Thought*. Albany: SUNY Press, 1994.
———. *The Genesis of Heidegger's Being and Time*. Berkeley: University of California Press, 1993.
Kisiel, Theodore, and Thomas Sheehan, eds. *Becoming Heidegger: On the Trail of His Early Occasional Writings, 1910–1927*, ed. Theodore Kisiel and Thomas Sheehan. Evanston: Northwestern University Press, 2007.
Kohak, Erazim. *The Embers and the Stars: A Philosophical Inquiry into the Moral Sense of Nature*. Chicago: University of Chicago Press, 1984.
Krell, David Farrell. *Daimon Life: Heidegger and Life-Philosophy*. Bloomington: Indiana University Press.
———. "Heidegger's Reading of Nietzsche: Confrontation and Encounter." *Journal of the British Society for Phenomenology* 14, no. 3 (October 1983).
Lampert, Lawrence. "Heidegger's Nietzsche Interpretation." *Man and World: An International Philosophical Review* 7, (November 1974): 353–78.
Leopold, Aldo. *A Sand County Almanac*. New York: Oxford University Press, 1949.
Levin, David Michael. *The Opening of Vision: Nihilism and the Postmodern Situation*. London: Routledge, 1985.
Lewis, C. S. *The Abolition of Man*. New York: HarperCollins, 1944.
Lewis, Michael. *Heidegger and the Place of Ethics*. New York: Continuum, 2005.
Lightbody, Brian. *The Problem of Naturalism: Analytic Perspectives, Continental Virtues*. New York: Lexington Books, 2013.
Llewelyn, John. *The Middle Voice of Ecological Conscience: A Chiasmic Reading of Responsibility in the Neighborhood of Levinas, Heidegger, and Others*. New York: St. Martin's, 1991.

Lovejoy, Arthur. *The Great Chain of Being*. New York: Harvard University Press, 1936.
Lowith, Karl. *Martin Heidegger and European Nihilism*, trans. Gary Steiner. New York: Columbia University Press, 1998.
———. *Nature, History, and Existentialism*, ed. Arnold Levison. Evanston: Northwestern University Press, 1966.
Luft, E. von der. "Sources of Nietzsche's 'God Is Dead!' and Its Meaning for Heidegger." *Journal of the History of Philosophy* 45 (1984): 263–76.
MacNeill, Will. *The Time of Life*. Albany: SUNY Press, 2007.
Maslow, Abraham. *The Farther Reaches of Human Nature*. New York: Penguin, 1993.
McWhorter, Ladelle, ed. *Heidegger and the Earth*. Kirksville: Jefferson University Press, 1992.
Merleau-Ponty, Maurice. *The Structure of Behavior*, trans. Alden Fisher. Boston: Beacon Press, 1963.
Moles, A. *Nietzsche's Philosophy of Nature and Cosmology*. New York: Peter Lang Publishing Co., 1990.
Morris, David. "Animals and Humans, Thinking and Nature." *Phenomenology and the Cognitive Sciences* 4 (2005): 49–72.
Moore, Gregory. *Nietzsche, Biology, Metaphor*. Cambridge: Cambridge University Press, 2002.
Moyle, Tristan. "Re-enchanting Nature: Human and Animal Life in Later Merleau-Ponty." *Journal of the British Society for Phenomenology* 38, no. 2 (May 2007).
Naess, Arne. *Ecology, Community, Lifestyle*. Cambridge: Cambridge University Press, 1989.
Nietzsche, Friedrich. *Basic Writings*, ed. and trans. Walter Kaufmann. New York: Modern Library Edition, 2000.
———. *Beyond Good and Evil*, trans. Walter Kaufmann. New York: Vintage Books, 1966.
———. *The Gay Science*, trans. Walter Kaufmann. New York: Vintage, 1974.
———. *The Will to Power*, ed. and trans. Walter Kaufmann. New York: Vintage, 1968.
———. *Thus Spoke Zarathustra*, trans. Graham Parkes. New York: Oxford University Press, 2004.
Norton, Bryan. "Environmental Ethics and Weak Anthropocentrism." *Environmental Ethics* 6 (1984): 133–38.
Oliver, Kelly. "Stopping the Anthropological Machine." *PhaenEx* 2, no. 2 (Fall/Winter 2007).
———. "Strange Kinship: Heidegger and Merleau-Ponty on Animals." *Epoche* 7, no. 1 (2008).
Parkes, Graham. *Composing the Soul: Reaches of Nietzsche's Psychology*. Chicago: University of Chicago Press, 1994.
———, ed., *Heidegger and Asian Thought*. Honolulu: University of Hawaii Press, 1987.

———. "Nature and the Human 'Redivinized': Mahayana Buddhist Themes in *Thus Spoke Zarathustra*." In *Nietzsche and the Gods*, ed. Weaver Santaniello. Albany: SUNY Press, 2001.

———. "Nietzsche's Environmental Philosophy: A Trans-European Perspective." *Environmental Ethics* 27, no. 1 (Spring 2005): 77–91.

———. "Staying Loyal to the Earth: Nietzsche as an Ecological Thinker." In *Nietzsche's Futures*, ed. John Lippitt. Harmondsworth: Macmillan, 1998.

Passmore, John. *Man's Responsibility for Nature: Ecological Problems and Western Traditions*. London: Duckworth, 1974.

Pearson, Keith Ansell. "Incorporation and Individuation: On Nietzsche's Use of Phenomenology for Life." *Journal of the British Society for Phenomenology* 39, no. 1 (2007): 61–89.

Polt, Richard, and Gregory Fried, eds. *A Companion to Heidegger's* Introduction to Metaphysics. New Haven: Yale University Press, 2001.

Regan, Tom. *The Case for Animal Rights*. Berkeley: University of California Press, 1983.

Reginster, Bernard. *The Affirmation of Life: Nietzsche on Overcoming Nihilism*. Cambridge: Harvard University Press, 2006.

Richards, Robert J. *The Romantic Conception of Life: Science and Philosophy in the Age of Goethe*. Chicago: University of Chicago Press, 2002.

Richardson, John. *Nietzsche's New Darwinism*. New York: Oxford University Press, 2004.

Rosen, Stanley. *Nihilism: A Philosophical Essay*. New Haven: Yale University Press, 1969.

———. *The Question of Being: A Reversal of Heidegger*. New Haven: Yale University Press, 1993.

Rue, Loyal. *By the Grace of Guile: The Role of Deception in Natural History and Human Affairs*. New York: Oxford University Press, 1994.

Sallis, John, ed. *Continental Philosophy in America: Prize Essays, Volume I*. Pittsburgh: Duquense University Press.

Schalow, Frank. "Essence and Ape: Heidegger and the Question of Evolutionary Theory." *American Catholic Philosophical Quarterly* 82, no. 3 (2008): 2008.

———. *The Incarnality of Being: The Earth, Animals, and the Body in Heidegger's Thought*. Albany: SUNY Press, 2006.

Scheler, Max. *Man's Place in Nature*, trans. Hans Meyerhoff. New York: The Noonday Press, 1961.

Schrift, Alan. *Nietzsche's French Legacy*. New York: Routledge, 1995.

———. "Nietzssche's Psycho-Genealogy: A Ludic Alternative to Heidegger's Reading of Nietzsche." *Journal of the British Society for Phenomenology* 14, no. 3 (October 1983).

Seidel, George J. "Heidegger: Philosopher for Ecologists?" *Man and World* 4 (1971): 93–99.

Sessions, George, and Bill Devall. *Deep Ecology: Living as if Nature Mattered*. Salt Lake City: Peregrine Smith Books, 1985.

Shaner, David Edward. "Beneath Nihilism: The Phenomenological Foundations of Meaning." *Personalist Forum* 3, 113–39.
Sheehan, Thoman. "A Paradigm Shift in Heidegger Research." *Continental Philosophy Review* 32, no. 2 (2001): 1–20.
———. "Nihilism: Heidegger/Junger/Aristotle." In *Phenomenology: Japanese and American Perspectives*, ed. Bert C. Hopkins. Dordrecht: Kluwer Academic Publishers, 1998, 273–316.
Shook, John R., and Paul Kurtz, eds., *The Future of Naturalism*. New York: Humanity Books, 2009.
Sommers, Tamler, and Alex Rosenberg. "Darwin's Nihilistic Idea: Evolution and the Meaninglessness of Life." *Biology and Philosophy* 18 (2003): 653–58.
Soule, Michael. *Reinventing Nature? Responses to Postmodern Deconstruction*. Washington, D.C.: Island Press, 1995.
Taminiaux, Jacques. *Heidegger and the Project of Fundamental Ontology*, trans. Michael Gendre. Albany: SUNY Press, 1991.
———. "On Heidegger's Interpretation of the Will to Power as Art." *New Nietzsche Studies* 3, nos. 1–2 (Winter 1999).
Taylor, Charles. *A Secular Age*. Cambridge: Belknap Press, 2008.
———"Heidegger, Language, and Ecology." In *Heidegger: A Critical Reader*, ed. Hubert Dreyfus. Cambridge: Blackwell, 1992.
Taylor, Paul W. *Respect for Nature: A Theory of Environmental Ethics*. Princeton: Princeton University Press, 1986.
Thiele, Leslie Paul. "Nature and Freedom: A Heideggerian Critique of Biocentric and Sociocentric Environmentalism." *Environmental Ethics* 17, no. 2 (Summer 1999): 171–90.
Thompson, Evan. *Mind in Life: Biology, Phenomenology, and the Sciences of Mind*. Cambridge: Harvard University Press, 2007.
———. "Planetary Thinking/planetary building: An essay on Martin Heidegger and Nishitani Keiji." *Philosophy East and West* 36, no. 3 (July 1986): 235–52.
Thompson, Iain. "Ontology and Ethics at the Intersection of Phenomenology and Environmental Philosophy." *Inquiry* 47 (2004): 380–412.
Toadvine, Ted. *Merleau-Ponty's Philosophy of Nature*. Evanston: Northwestern University Press, 2009.
Uexküll, Jakob von. *A Foray into the World or Animals and Humans, with a Theory of Meaning*, trans. Joseph D. O'Neil. Minneapolis: University of Minnesota Press, 2010.
Van Buren, John, ed. *Supplements: From the Earliest Essays to* Being and Time *and Beyond*. Albany: SUNY Press, 2002.
———. *The Young Heidegger: Rumors of the Hidden King*. Bloomington: Indiana University Press, 1994.
VanDeVeer, Donald. "Interspecific Justice." *Inquiry: An Interdisciplinary Journal of Philosophy* 24, nos. 1–4 (1979): 55–79.
VanDeVeer, Donald, and Christine Pierce, eds. *The Environmental Ethics and Policy Book*. Toronto: Nelson Thomson Learning, 2003.

Varner, Gary E. "The Schopenhauerian Challenge in Environmental Ethics." *Environmental Ethics* 7 (1985): 209–30.
Vogel, Steven. *Against Nature: the Concept of Nature in Critical Theory.* Albany: SUNY Press, 1996.
Vogel, Lawrence. "Hans Jonas's Diagnosis of Nihilism: The Case of Heidegger." *International Journal of Philsophical Studies* 3, no. 1 (March 1995): 55–72.
Westra, Laura. "Let It Be: Heidegger and Future Generations." *Environmental Ethics* 7, no. 4 (Winter 1985): 341–50.
Wilkerson, Dale Allen. "The Root of Heidegger's Concern for the Earth at the Consummation of Metaphysics: The Nietzsche Lectures." *Cosmos and History: The Journal of Natural and Social Philosophy* 1, no. 2 (2005): 27–34.
White, A. "Nietzschean Nihilism: A Typology." *International Studies in Philosophy* 19 (1987): 29–44.
White, Dick. "Heidegger on Nietzsche on the Question of Value." In *Postmodernism and Continental Philosophy*, ed. Hugh Silverman. Albany: SUNY Press, 1988.
White, Jr., Lynn, "The Historical Roots of Our Ecological Crisis." In *The Environmental Ethics and Policy Book*, ed. Donald VanDeVeer and Christine Pierce. Toronto: Nelson Thomson Learning, 2003, 52–58.
Wilber, Ken. *Sex, Ecology, Spirituality: The Spirit of Evolution.* Boston: Shambhala, 2000.
Young, Julian. "Being and Value: Heidegger Contra Nietzsche." *International Studies in Philosophy* 27, no. 3 (1985): 100.
———. *Heidegger's Later Philosophy.* New York: Cambridge University Press, 2002.
Zimmerman, Michael E. *Contesting Earth's Future: Radical Ecology and Postmodernity.* Berkeley: University of California Press, 1994.
———. *Heidegger's Confrontation with Modernity.* Bloomington: Indiana University Press, 1990.
———. "Heidegger, Buddhism, and Deep Ecology." *The Cambridge Companion to Heidegger*, ed. Charles Guignon. New York: Cambridge University Press, 1993.
———. "Nietzsche and Ecology: A Critical Inquiry." In *Reading Nietzsche at the Margins*, ed. Steven V. Hicks and Alan Rosenberg: Purdue University Press, 2007.
———. "Rethinking the Heidegger—Deep Ecology Relationship." *Environmental Ethics* 15 (1993): 95–224.
———. "Toward a Heideggerian Ethos for Radical Environmentalism." *Environmental Ethics* 6, no. 2 (1983): 106.
———, and Sean Esbjörn-Hargens. *Integral Ecology: Uniting Multiple Perspectives on the Natural World.* New York: Integral Books, 2009.

Index

Abram, David, 20
Acampora, Ralph, 175, 217–28
Agamben, Giorgio, 34, 83, 96, 100
aletheia, 120, 121, 127
alterity, 20, 78, 83, 96, 222–23, 233
animality, 6, 7, 12, 33–43 passim, 62, 83–101 passim, 119, 167, 170, 183
anthropocentrism, 18–24, 77, 81, 108, 138–39, 170, 218, 226, 228
anthropomorphism, 37, 179, 181, 191, 195–96
anxiety, 28, 51–56 passim, 75–76, 84–85, 116, 132
Aristotle, 6, 11, 14–18, 29–48 passim, 53–55, 62–63, 69, 78, 83–84, 88, 92, 95, 99, 104, 118–30 passim, 143, 160, 173, 198–99, 205
art, 69–70, 76, 109–10, 113–22 passim, 125–26, 130, 133–35, 142, 155, 168, 178, 219; see *Origin of the Work of Art*
authenticity, 47, 55–56, 76, 87, 128, 132–33, 137, 153, 156, 168
autopoiesis, 197–201
axiology, 11, 22, 28, 220

Baer, Karl von, 44, 64, 95
Bambach, Charles, 96, 109, 142, 153, 154
Baumler, Alfred, 109, 142, 156
being, 5, 8, 11–28 passim, 31, 34–38, 40, 47–81 passim, 84–85, 88–89, 92, 95–101 passim, 104, 108, 112–15, 118–62 passim, 166–72, 194–96, 198–200; *see* ontology; metaphysics
Being and Time (Heidegger), 7, 11, 15, 17, 21–33 passim, 37–38, 43–78 passim, 83–92, passim, 99, 101, 107–21 passim, 127–55 passim, 161, 166, 205
being-in-the-world, 21–27 passim, 38–62 passim, 68, 72, 77, 78, 88, 101, 133
Bildungstrieb, 177, 179
biocentrism, 7, 9, 107, 214–20 passim, 228
biology, 3–9, 12–14, 20, 25–26, 29–50 passim, 59–66 passim, 75, 78, 83–84, 87, 89–90, 94–95, 100, 104, 118, 141, 159, 172, 173–218 passim, 223–24, 228; Nietzsche's philosophical, 173–218 passim, 223–24, 228
body, 20, 42, 61, 71, 90–91, 94, 98, 118–23 passim, 155, 170, 185, 195, 198, 203, 214, 217, 226
boredom, 83–87
Buchanan, Brett, 20, 42, 63, 88, 94, 95

Caputo, John, 96, 109, 119, 142–52 passim
consciousness, 18–30 passim, 52, 64–65, 72, 94, 102–103, 122,

consciousness *(continued)*
    127, 138, 145, 153, 164–65, 170, 177–78, 187, 203–205, 212–13, 222, 225–26
cosmos, 2, 15, 45, 72, 76–78, 82, 91, 98, 102, 138, 150, 153, 163, 165, 194, 199, 224
creation, 2, 15, 138, 150

Darwin, Charles, 5, 13, 18, 25, 29, 33, 84, 93–95, 111, 138, 175–88 passim, 202–203, 211, 233
Dasein, 15, 22, 28, 38–78 passim, 82–88 passim, 92, 128–37 passim, 143–55 passim, 161, 168, 194, 196
Davis, Bret, 137, 145–46, 155, 161–62, 168
deep ecology, 16–19, 107, 219–23, 232
Derrida, Jacques, 12, 34, 83, 96, 100
Descartes, Rene, 23, 28, 54, 121, 142, 156, 159, 170
Dennett, Daniel, 93, 176–91 passim, 202
Drenthen, Martin: 221–25 passim
Dreyfus, Hubert: 48, 52, 66–78 passim, 132
Drives, 26, 38, 88, 90, 100, 123, 131, 158–62 passim, 167–93 passim, 201–14 passim, 221–30 passim
dualism, 4, 13, 15, 18, 98–104 passim, 147, 150, 156, 159–60, 190–98 passim, 213, 218, 220, 222

ecology, 7, 16, 18–19, 25–29, 48, 78, 107, 110–12, 118, 138, 219–23 passim, 230, 232
earth: 7, 11, 16–18, 20, 30, 35, 47–48, 57–58, 76–78, 82, 91, 98–127 passim, 131–36 passim, 144–47, 152–53, 163–64, 169, 172, 193, 222, 224; Heidegger's concept of, 108–33 passim

ecocentrism, 18, 71
ecophenomenology, 14, 26–27, 82
ecosystem, 19, 108–12, 219–20, 232
entelechy, 41, 90, 122, 126
environmental ethics, 1–9 passim, 13–14, 18–22, 31, 71, 81–82, 105, 108, 118, 133, 138–40, 153, 169, 174, 215, 217–22, 227–34
Ereignis, 92, 115, 144
Esbjorn-Hargens, Sean, 110–11, 220, 230–31
evolution, 5–9 12–13, 25, 29–31, 71, 78–79, 81, 83, 87, 92–97 passim, 102, 111, 102, 156, 158, 160, 172, 173–219 passim, 224, 228–29, 233; Heidegger's view of, 93–96; Nietzsche's view of, 173–219 passim, 224
existentialism, 4, 15, 194

factical life, 11, 37–39, 41, 55, 85, 152
Foltz, Bruce, 5, 18, 21, 48, 66–82 passim, 98–100, 132, 169
form, 2, 40–41, 92, 95, 113–14, 117–18, 122–26, 134–35, 158–62, 167–70, 182–92 passim, 197–200, 203–204, 209, 211, 217–19
fourfold, 17, 30, 102, 107, 109, 133, 135–37, 139, 142, 144
fundamental ontology, 20, 34, 39, 48, 54, 58, 62–63, 65–66, 71–73, 77–78, 81, 88, 109, 116, 139, 141–43, 150, 155
Fundamental Concepts of Metaphysics (Heidegger), 6–7, 20, 33–38, 47–48, 62–66, 83–85, 88, 91–92, 96, 99–100, 150–51, 223–24

Gelassenheit, 8, 17, 31, 107, 123, 133, 135, 137–38, 142–47 passim, 153, 221–22, 227
Gestell, 17, 112–13, 128, 130–32, 135–37, 142–46, 160, 168

great chain of being, 9, 15, 76, 88, 96, 101, 177, 191, 194, 215, 217, 234
Grene, Marjorie, 28–29, 83, 95
Griffin, David Ray, 14, 218, 230, 232

Haar, Michel, 11, 58, 108–10, 118, 132–45 passim
Haeckel, Ernst, 93, 110–11, 184
Hallman, Max, 217–28 passim
Hegel, G.W.F., 101–102, 121, 145, 151, 159–60, 164, 224
Heidegger, Martin: appropriation of Uexküll, 35–37, 88, 90, 94–95, 98; *Being and Time,* 7, 11, 15, 17, 21–33 passim, 37–38, 43–78 passim, 83–92, passim, 99, 101, 107–21 passim, 127–55 passim, 161, 166, 205; concept of life, 62–66, 81–84, 88–105, 191, 193, 205; early Aristotle lecture courses, 37–45; *Fundamental Concepts of Metaphysics,* 6–7, 20, 33–38, 47–48, 62–66, 83–85, 88, 91–92, 96, 99–100, 150–51, 223–24; interpretation of Nietzsche, 154–72; *Introduction to Metaphysics,* 11, 38, 88, 119–22, 143, 146–54; later philosophy of nature, 107–40; *Origin of the Work of Art,* 11, 98–127 passim; *Question Concerning Technology,* 128–33; relationship to environmental philosophy, 14–31, 217, 221–23, 233; view of evolution, 93–96; view of nihilism, 54–57, 84–87, 143–54; *What is Metaphysics?,* 83, 94; *Will to Power as Art,* 155; *see* earth; *physis*; poetic dwelling; *Gelassenheit*; *Gestell*
hermeneutics, 23, 38, 94, 102–103
hierarchy, 5, 15, 19, 43, 83, 88, 91, 95–100, 170, 177, 191, 194, 199, 212, 215, 218–22, 227–31
historical decline, 122, 142, 146, 149, 152, 170

history of being, 7, 31, 48, 78, 85, 96, 107–109 119, 128, 141–46, 150–56, 160–61
Holderlin, Friedrich, 122–23, 132
Humanism, 7, 18, 24, 28, 56, 78, 119, 131, 136, 138, 146, 157, 160, 218
Husserl, Edmund, 3, 19–28 passim, 36–37, 47, 62–65, 82, 150, 157

inauthenticity, 47, 55–56, 87, 128, 132, 137, 168
intentionality, 3, 6, 9, 13, 21, 25–26, 29, 42, 47, 50, 60, 65, 67, 72, 75, 101, 104–105, 107, 115, 132, 167, 193–94, 197–98, 201–205 passim, 210, 222
interiority, 29, 42, 77, 91–92, 119, 139, 177, 198, 222, 230
intersubjectivity, 42, 91, 231
Introduction to Metaphysics (Heidegger), 11, 38, 88, 119–22, 143, 146–54

Jonas, Hans, 8, 15–18, 28–29, 39, 76, 82–83, 95, 98, 138, 153, 160, 173, 180, 193–200, 217, 223
Junger, Ernst, 123, 142, 150, 156

Kant, Immanuel, 12–13, 22, 28, 30, 33, 35–36, 50, 54, 55, 62–63, 73, 78, 83–84, 101–104, 150, 170, 188–92, 199–202, 208, 222, 233
Kaufmann, Walter, 154–60 passim, 229, 233
Kohak, Erazim, 19–20, 230
Krell, David Farrell, 20, 34, 38–41, 65–66, 81, 83, 90, 93, 96–101, 155, 161, 166, 170

Lamarck, Jean Baptiste, 177, 185, 206–207
Lampert, Laurence, 154, 157, 163–64
Lebensphilosophie, 20, 62, 64–65, 93, 222

Leiter, Brian: 175–76, 224
Leopold, Aldo, 86, 138
Lowith, Karl, 15, 18, 21, 28, 82, 138, 153

machination, 7, 107, 114, 128–29
materialism, 3, 17, 25, 29, 81, 93, 95, 111, 120, 124, 160, 173, 179–80, 194–96, 207
matter, 2–5, 24, 40, 91, 104, 110, 113–14, 118, 122–25, 134–35, 158, 160, 177, 179, 186, 190–92, 194–202 passim, 209, 223
McNeill, William, 83, 90, 96, 99
mechanism, 1–2, 5–6, 8, 13, 29–37 passim, 63, 81, 84, 88, 90–94, 102–104, 112, 114, 158, 160, 173–95 passim, 200–202, 206–209 210, 212, 215, 220, 223, 226, 228
Merleau-Ponty, Maurice, 19–20, 192, 210
metaphysics, 3, 5, 8–9 12–18 passim, 28, 30, 36, 54, 62, 68, 70, 77, 85, 92, 94, 99, 104, 107, 114, 118–23, 127–29, 141–46, 149, 151, 153–54, 157–58, 160, 166, 171–72, 174, 176, 180, 194, 198, 206–207, 211, 217, 224–26; *see* ontology; being
modernity, 1, 19, 24, 55–56, 86–87, 127, 129, 142, 147–48, 151–52, 164, 167, 182, 233
Moore, Gregory, 93, 176–88 passim, 204, 211
motion, 33, 39–41, 84–85, 104, 120, 122–24, 190, 192, 209

Naess, Arne, 16, 232
naturalism: 1–16 passim, 21, 24–38 passim, 45, 64, 71, 78–79 82–83, 104, 119, 136–37, 141, 153, 160, 172–81 passim, 188, 192–94, 200, 215, 217, 222–23, 228 scientific; phenomenological; critique of; nonreductive

naturalization, 1, 6–9 12, 14, 26, 31, 82, 123, 141, 156, 164, 181, 201–202, 209, 215
neo-Darwinism, 180, 181, 194, 200, 206–208, 211
neo-Kantianism, 22, 65
Nietzsche, Friedrich, 1, 5–6, 8–9 12–14, 16, 22, 28, 30–31, 55–56, 65, 76, 82–83, 86, 94, 98, 114, 119, 121, 123, 140–94 passim, 199–233 passim; and environmental ethics: 217–33; Heidegger's interpretation of: 154–72; naturalism of: 174–78; philosophy of biology: 178–94, 202–15; value theory of: 208–11; view of nihilism: 141–43, 162–67
nihilism, 4–9 28, 31, 48, 54–56, 65, 76, 81–87 passim, 94–95, 109, 122–23, 127–28, 135, 137, 141–66 passim, 169, 175–76, 178, 199–200, 206, 208, 212–18 passim, 222–23, 229, 234; Heidegger's view of, 54–57, 84–87, 143–54; Nietzsche's view of: 141–43, 162–67
nonanthropocentrism, 7, 107, 138
nothingness, 56, 83, 116, 141

Oliver, Kelly, 96–97
ontology, 6, 15, 17, 19–20, 29, 33–58 passim, 62–73 passim, 78, 81, 84, 88, 91, 94, 101–104, 109, 114, 116, 139, 141–43, 150, 155, 181, 191–96, 199–200, 206, 215, 218, 220; *see* being; metaphysics
organism, 6, 9, 13–14, 20, 28–30, 35–36, 41–42, 64, 81, 88–90, 94–95, 103–104, 110–12, 119, 139, 152, 162, 179–89, 194, 197–217 passim, 227, 230
Origin of the Work of Art (Heidegger), 11, 109

panpsychism, 170, 177, 181, 186–87, 199

Parkes, Graham, 76, 110, 159, 170, 175, 205, 217, 221, 224–28
phenomenology, 1–30 passim, 36–37, 48, 50, 61–66, 77–78, 82–83, 88–95, 103, 124, 143, 173, 193–94, 198, 200, 223; eco-: 14, 19, 26–27, 82; naturalized: 7, 9, 26, 82; transcendental: 12, 28, 30, 39, 62, 65, 78, 82–83, 39
physis, 5, 7, 11, 17, 30, 35, 37–38, 47–48, 58, 66, 70, 76, 78, 81–85, 91, 97–109 passim, 116–35 passim, 155–51 passim, 155, 168, 219
Plato, 125, 154, 174, 211
poetic dwelling, 5, 7, 107, 109, 133–34, 136–37, 139
poiesis, 130, 133–35, 155, 197–99, 201
positivism, 37, 91, 149, 153, 160, 164, 174, 225
presence-at-hand, 21, 40, 53, 58–62, 68–69, 71–72, 75
principle of continuity, 89, 98, 101
psychology, 23, 25, 50, 87, 100, 141, 143, 150–51, 156–58, 181, 185, 193, 199, 202, 205, 212, 226, 228

Question Concerning Technology (Heidegger), 128–33

readiness-to-hand, 51, 58–61, 68, 71–72
realism, 9, 30, 73–75, 78, 82, 217
Richards, Robert, 103–104, 188–91, 211
Richardson, John, 174–86 passim, 201–14 passim, 218, 224
Rickert, Heinrich: 22, 65, 166
romanticism, 58, 69, 111

Sartre, Jean-Paul, 28, 75–76
scala natura, 9, 15, 43, 96, 211, 217
Schacht, Richard, 175–76
Schalow, Frank, 81, 93–94

Scheler, Max, 29, 65, 86, 193
Schelling, F.W.S., 65, 101–102, 159, 188, 190–91, 224
science, 1–9 passim, 17–21, 25–26, 30, 34–37, 59–64, 68–78 passim, 82, 84, 89–96 passim, 102–104, 112, 114, 118, 127, 129, 132, 139–40, 149, 153, 157, 159, 172–80, 188–204 passim, 213–24 passim, 229, 223
Schopenhauer, Arthur, 159–60
Spencer, Herbert, 94, 178, 182–83, 187
Spengler, Oswald, 65, 86, 142, 150–51, 156
spirit, 65, 86–87, 100, 102, 113, 147–48, 151, 163–64, 177, 179, 187, 195
subjectivity, 16, 18, 23–24, 28, 30, 50, 91, 142, 150, 157, 160, 177, 192, 204, 209, 232
sublime, 73, 78, 87, 101, 102, 184, 191, 221–22

Taminiaux, Jacques, 54, 116, 136, 155, 161
technology, 5, 7, 14–17, 20, 24–25, 78, 86, 95, 107, 113–14, 121–22, 127–38 passim, 142–53 passim, 160–61, 167–68, 223, 229, 233
teleology, 33, 103–104, 173–75, 181–92 passim, 199–205 passim, 224
temporality, 49–53 passim, 57, 77, 92, 97, 136, 152, 155, 165
Thompson, Evan, 8, 13, 15, 173, 190, 192–94, 197, 200, 210, 223
Toadvine, Ted, 26–27, 29, 91, 222
truth, 22–23, 27, 44, 72–73, 109, 115, 117, 120, 129, 133, 145, 149, 163–64, 166–67, 170, 174; see *aletheia*

Uexküll, Jakob von, 6, 14, 20, 21, 29–44 passim, 62, 64, 88, 90,

Uexküll, Jakob von *(continued)*
  94–95, 98, 139, 143, 173, 179, 185, 210
Umwelt, 20–21, 30, 35, 38, 40, 42–43, 45, 60, 62, 64, 81, 88, 90, 101, 147, 197, 201, 210

value, 1–9 12–31 passim, 55–57, 665, 68, 76, 81–82, 94, 96, 121, 138, 141, 143, 147, 150–51, 156–57, 160–87 passim, 193–96, 200–34 passim
vitalism, 8, 41, 63, 88, 90, 93, 98, 159–60, 176–80, 187, 192, 204, 206, 224

"What is Metaphysics?" (Heidegger), 83, 94
Whitehead, Alfred North, 4, 13, 77, 126, 199, 207–208

will to power, 121, 140–43, 150–86 passim, 193, 202, 205–206, 208, 219, 224–27
Will to Power as Art (Heidegger), 155
world, 3–13 passim, 19–78 passim, 87–101 passim, 108–18 passim, 130–33, 136–231 passim; in *Being and Time*: 47–78; and earth: 108–18; in *Fundamental Concepts of Metaphysics*: 87–101; and *Gestell*: 130–33; *see* being-in-the world
worldview, 2, 3, 21, 151, 163–64, 182, 201, 215, 219, 223

Young, Julian, 115, 132–36, 164–69

Zimmerman, Michael E., 5, 18–24 passim, 76, 107, 110–11, 114, 150–51, 168, 171–72, 177, 218, 220, 230, 232

www.ingramcontent.com/pod-product-compliance
Ingram Content Group UK Ltd.
Pitfield, Milton Keynes, MK11 3LW, UK
UKHW041916140426
5217IPUK00013B/184